4 R005-829 33+ 89 590
KKR - Šimandl
69

Lecture Notes in Control and Information Sciences

Edited by M. Thoma and A. Wyner

124

D1666064

A. A. Bahnasawi,
M. S. Mahmoud

58447
Vysoká škola strojní a elektrotechnická v Plzni knihovna
VYŘAZENO
5. října 1990

Control of Partially-Known Dynamical Systems

VYŘAZENO
ZÁPADOČESKÁ UNIVERZITA v Plzni
Technicko – ekonomická knihovna
3143063293
NENIČ MNE !

Springer-Verlag
Berlin Heidelberg New York
London Paris Tokyo Hong Kong

Series Editors
M. Thoma · A. Wyner

Advisory Board
L. D. Davisson · A. G. J. MacFarlane · H. Kwakernaak
J. L. Massey · Ya Z. Tsypkin · A. J. Viterbi

Authors
Dr. Ahmed A. Bahnasawi
Electronics and Communication Engineering Department
Cairo University
Giza
Egypt

Prof. Magdi S. Mahmoud
Techno-Economics Division
Kuwait Institute for Scientific Research
P.O. Box 24 88 5
13109-Safat
Kuwait

ISBN 3-540-51144-X Springer-Verlag Berlin Heidelberg New York
ISBN 0-387-51144-X Springer-Verlag New York Berlin Heidelberg
ISBN 13: 9783540511441

Library of Congress Cataloging in Publication Data
Bahnasawi, A. A. (Ahmed A.)
Control of partially-known dynamical systems / A. A. Bahnasawi, M. S. Mahmoud.
(Lecture notes in control and information sciences ; 124)
Bibliography: p.
ISBN 0-387-51144-X (U. S. : alk. paper)
1. Automatic control. 2. Adaptive control systems. 3. Feedback control systems. 4. Dynamics.
I. Mahmoud, Magdi S. II. Title. III. Series.
TJ 211.B 33 1989
629.8--dc20 89-10085

This work is subject to copyright. All rights are reserved, whether the whole or part of the material is concerned, specifically the rights of translation, reprinting, re-use of illustrations, recitation, broadcasting, reproduction on microfilms or in other ways, and storage in data banks. Duplication of this publication or parts thereof is only permitted under the provisions of the German Copyright Law of September 9, 1965, in its version of June 24, 1985, and a copyright fee must always be paid. Violations fall under the prosecution act of the German Copyright Law.

© Springer-Verlag Berlin, Heidelberg 1989
Printed in Germany
The use of registered names, trademarks, etc. in this publication does not imply, even in the absence of a specific statement, that such names are exempt from the relevant protective laws and regulations and therefore free for general use.

Offsetprinting: Mercedes-Druck, Berlin
Binding: B. Helm, Berlin
2161/3020-543210 – Printed on acid-free paper.

TO OUR WIVES

Somaya (A A B) and Salwa (M S M)

For their encouragement,
understanding and support.

ABOUT THE AUTHORS

Ahmed A. Bahnasawi was born in Cairo, EGYPT in 1958. He received the BSEE degree (Distinction with Honours) in Electronic Engineering, the MSEE degree in Control Engineering, and the Ph. D. degree in Systems Engineering, all from Cairo University in 1981, 1985, 1988, respectively. Since 1981 he has been with Cairo University, first as an assistant Tutor, then as an Instructor at the Electronics and Communications Engineering Department and he is currently an assistant Professor. His current research interests include stabilization of uncertain systems, adaptive control and problems of large scale systems.

Magdi S. Mahmoud (S'73-M'75-SM'83) was born in Cairo, EGYPT in 1948. He received the BSEE degree (Honours) in Communications Engineering, the MSEE degree in Electronic Engineering, and the Ph. D. degree in Systems Engineering, all from Cairo University in 1968, 1972, 1974, respectively.

He has served on the faculties of Al-Azhar University (EGYPT), Cairo University (EGYPT), University of Manchester (UK), Pittsburgh University (USA), Case Western Reserve University (USA) and Kuwait University (KUWAIT). Since 1984 he has been Professor of Control Systems Engineering at Cairo University. He is the principal author of three books, the coauthor of three textbooks and the author/coauthor of more than 130 technical articles. His research, teaching and consulting activities are in the areas of :(a) control problems of complex systems, (b) computer control systems and (c) systems engineering simulation and management.

Dr Mahmoud was the recipient of the 1978, 1986 SCIENCE STATE INCENTIVE PRIZES for outstanding research in engineering, EGYPT and of the 1986 ABDULHAMED SHOWMAN PRIZE for young arab scientists in engineering sciences, JORDAN. He also holds the STATE MEDAL OF SCIENCE AND ARTS (first class), EGYPT. He is listed in the 1979 edition of WHO'S WHO IN TECHNOLOGY TODAY, Pittsburgh, USA. He is the Vice-Chairman of the IFAC-SECOM Working Group on Large-Scale Systems Methodology and Applications and an Associate Editor of Large Scale Systems.

Dr Mahmoud is a fellow of the IEE, a senior member of the IEEE, a member of Sigma Xi, the CEI (UK), the Egyptian Engineers Society (EGYPT), and the Kuwait Engineers Society (KUWAIT) and is a registered Consultant Engineer of Information Engineering and Systems in Egypt.

ACKNOWLEDGEMENTS

The research reported in this monograph has been conducted in the Electronics and Communications Engineering Department of Cairo University - EGYPT, chared by Professor E. A. Talkhan, with the active participation of Professor A. Y. Belal, Head of the National Institute of Telecommunications (NIT).

We are highly grateful to our colleagues Drs. S. Z. Eid, M. F. Hassan and M. G. Darwish, for reading different parts of the manuscript and for valuable suggestions.

The computational facilities provided by the computing center of the Faculty of Engineering, Cairo University were the key figures to our continuing research and we appreciate the great help.

We welcome any constructive criticism of the monograph and will be grateful for any appraisal by the readers.

A. A. Bahnasawi - M. S. Mahmoud

EGYPT - April, 1988.

TABLE OF CONTENTS

Page

CHAPTER 1

INTRODUCTION

1.1 Background

Scientists and engineers are often confronted with analysis, design and synthesis of real-life problems. The first step in such studies is the development of a "mathematical models" which can be a substitute for the real problem. In control engineering, model building from measurements on a dynamical system is known as identification, and has enjoyed a sustained boom as a research topic for a decade and a half. It is worth mentioning that the identification is the first important task before the implementation of any procedure since the obtained results will be critically dependent upon the validity of the model. The accuracy of such an identification could be measured by the difference between the output of the real system and that of the model [1,2].

In many industrial and process control applications, the use of high performance control systems is very desirable [1,2], Usually, the plant parameters are poorly known or vary during normal operation resulting in some degree of uncertainty in the mathematical models governing such physical systems. In general, these uncertainties may be parameters, constant or time varying which are known or imperfectly known In addition, there may be unknown or imperfectly known inputs as well as measurement noises. Consequently, the problem of designing feedback controls for dynamic systems subject to internal uncertainty and/or external disturbance is of interest in its own right. Furthermore, the control policies should be simple to compute and realize while giving satisfactory performance. Complexity in control schemes usually imposes a cost in terms of reliability and limit their practicality [1-4].

In control engineering literature, the stabilization problem of systems with uncertain parameters has been treated in several categories according to different assumptions and approaches [3-9]. There are basically three classes of categories:

1- The Stochastic Approach: where a priori statistical characterization of the uncertainties in the system dynamics and of the disturbances which impinge on the system are available. Hence, a control that minimizes the expected value of some performance index is required, as in the case of stochastic optimal control [3]. However, this approach needs the knowledge of a distribution. After complicated computations, only the expected value of the performance index is minimized leaving the possibility of occasional bad behaviour of the system dynamics unchecked [7]. If the statistical property of the uncertain parameters cannot be assumed beforehand but can be identified in the course of dynamic process, the idea of adaptive or learning is used [1,2,4,5]. But on the other hand the procedure of adaptive control is usually complicated, expensive and impractical.

2- The Sensitivity Approach: which is based on the assumption that the parameter uncertainty is small in order to obtain first order perturbation equation and hence design a feedback controller that minimizes the performance or trajectory sensitivity [6]. If the uncertain parameters vary in a finite domain but the optimal control can be written as a Taylor series in the uncertain parameters, then the system is called optimally adaptive [7]; the first-order approximation of such a system is called the optimally sensitive system [8]. However, this approach is limited due to the assumption of small perturbation. For large uncertainty, it is needed to assume that the control and some related functions be analytic in the uncertain parameters, which is not easy to justify in general [9].

3- The Minimax Approach: where the variation of the uncertain elements is assumed to be bounded. In this case, the uncertain dynamical system is formulated by contingent differential equations and the asymptotic stability can be guaranteed via generalized dynamical systems [10-12]. It is interesting to note that the Minimax approach gives the minimum cost but, except in very restricted conditions, it ends up with a nonlinear complicated control law even in a linear regulator problem [9-11].

The research work in this book is mainly focused on two-phases : (i) reduced-order adaptive control schemes and (ii) deterministic stabilizing control for uncertain dynamical systems. In the following, a brief survey concerning these topics of research will be given.

I-Adaptive control:

The idea of adaptive control has its origin in the early days of control. However, it was not until the 1950s that serious attempts were made to design practical adaptive control systems [13]. The initial attempts were hampered by two principle difficulties : lack of suitable computer technology for implementation and absence of adequate supporting theory. With improvements in computer technology, it became feasible to experiment with various strategies for adaptive control. The methods ranged from simple gain adjustment procedures to sophisticated algorithms which attempted to achieve optimal regulation in a stochastic environments [1,4,14]. This culminated in several successful experiments involving adaptive control.

All existing adaptive control algorithms can be categorized as either direct or indirect. In direct or model reference adaptive control [15-17], the controller parameters are updated directly to force the unknown plant to behave asymptotically like a chosen reference model. Global stability of the regulated system has been proved subject to some restrictive assumptions on the plant. Specifically, the plant must be minimum phase and certain a priori information is required such as the relative degree of the plant and the sign of the gain [15-17].

Indirect adaptive control is applicable to nonminimum as well as minimum-phase systems, and only an upper bound on the system order is required a priori. The system parameters are identified first, then the control law is updated using the most recent parameter estimates. However, a well known problem with this scheme is that the estimated system model may not be controllable or stabilizable for certain parameter estimates, which are referred to as singular points. To avoid having the parameter estimates converge to a singular points, one can use a persistently exciting input to force the parameter estimates to converge to the true parameter values. This approach has been developed by a number of individuals; see for example [18-23].

It is worth mentioning that most of the adaptive control algorithms are designed with the assumption that the plant dynamics are exactly those of one member of a specified class of models. It is then natural to ask how the adaptive control system will behave when,as is inevitable in practice, the true plant is not perfectly described by any model of the given class. If the stability of the adaptive control system is guaranteed, provided only that the modeling error is sufficiently small in some sense, then one can say that the adaptive control algorithm is robust. To this end, it is clear that robust stability is very important for the practical applicability of adaptive control algorithms [24].

Unfortunately, a stable adaptive control algorithm is not necessarily robustly stable [25,26]. The reason is that the modeling error signal (e.g. unmodeled dynamics) appears as a disturbance in the adaptive law and may cause the divergence of the adaptive process. The fact that the disturbance is correlated with the plant input and output signals and, in addition, is of the same order of magnitude, is part of the complexity of the robustness problem [24].

As a first step towards robustness results, the stability of adaptive control systems in the presence of bounded external disturbances has been investigated by several authors [27-31]. These investigations were prompted by observations (e.g. see [28]) showing that a bounded external disturbance, even an asymptotically vanishing one, can cause the divergence of the adaptive process, and thereby instability. To prevent the latter, three main approaches have been made [24] :

(a) In the first approach [27,28,30], a dead zone is used in the adaptive law so that adaptation takes place only when the identification error exceeds a certain threshold. If the disturbance is bounded below this threshold, then it can be shown that the adaptation is always in the "right" direction and closed-loop system stability is achieved. In order to choose the size of the dead zone appropriately, a bound on the disturbance must be known.

(b) In the second approach [27,29], a modification of the adaptive law is used, which comes into operation only when the norm of the estimated controller parameters exceeds a certain value and has the effect that the parameter estimates remain bounded for all time. Closed-loop system stability is thus obtained in the presence of bounded disturbances of arbitrary, unknown size. In this case, a bound on the norm of the desired (unknown) controller parameters must be known.

(c) In the third approach, a σ-modification, i.e., an adaptive law with the extra term $-\sigma\theta$, $\sigma > 0$ is suggested and analyzed in [32]. Again, if the disturbance is known to be bounded, closed-loop system stability is obtained.

In summary, we point out that the basic idea, in all modifications suggested above, is to prevent instability by eliminating the pure integral action of the adaptive laws and to guarantee boundedness of all signals in the adaptive loop.

In the case when unmodeled dynamics are present, global stability cannot be guaranteed by simply eliminating the pure integral action of the adaptive laws[33]. As mentioned above, the unmodeled dynamics act as an external disturbance in the adaptive scheme and hence can no longer be assumed to be bounded. Despite this difficulty, however, several local results have been obtained in the literature for adaptive schemes applied to plants whose modeled parts are minimum phase and of relative degree one and whose unmodeled parts are due to fast and stable parasitics. In [34,35] it is shown that the σ-modification guarantees the existence of a "large" region of attraction from which all signals are bounded and the tracking error converges to a "small" residual set provided that the amplitude and frequency content of the reference input signal is away from the parasitic range. In the absence of parasitics, however, the residual tracking error may not be zero. Results related to this can be found in [33,36]

II-<u>Deterministic Uncertain Dynamical Systems</u>:

During recent years, a number of papers have appeared which deal with the design of stabilizing controllers for uncertain dynamical systems [9,12,37-52]. In these papers, the so-called uncertain dynamical systems are typically described by differential equations which contain parameters whose values are imprecisely known. In contrast to the stochastic control set-up, no assumptions are made concerning the statistics of the uncertain parameters in question. Instead, only a bound on the parameter variations is taken as given and the objective is to find a class of controllers which guarantees stable operation for all possible variations of the uncertain quantities.

Roughly speaking, the results to date fall into two categories. There are those results which might appropriately be termed structural in nature; e.g. see [38-42]. By this we mean that the uncertainty cannot enter arbitrarily into the state equations; certain preconditions must be met regarding the locations of the uncertainty within the system description. Such conditions are often referred to as matching assumptions. It is interesting to note that uncertainties in this situation can be tolerated with an arbitrarily large prescribed bound. A second body of results [9,37] might appropriately be termed nonstructural in nature. Instead of imposing matching assumptions on the system, the authors in [9,37] permit more general uncertainties at the expense of "sufficient smallness" assumptions on the allowable sizes of the uncertainties. However, the input matrix cannot be uncertain.

The results in [12,38,39] show that under the validity of matching conditions there exist nonlinear feedback control laws which guarantee asymptotic stability in the Lyapunov sense. The control laws typically require switching of the control signal values on hyperplanes in the state space, so the differential equations which govern the feedback system generally possess discontinuous right-hand sides. These equations can be treated mathematically by the theory of generalized dynamic systems [10,43], but the reported controllers are often difficult to implement.

If the feedback controller is restricted to be continuous in the state and time, the weaker property of uniform ultimate boundedness irrespective of uncertainty in input signals and parameter values may still be obtained [40,42]. The closed-loop system state will then reach, and remain within, a bounded set containing the zero state for all sufficiently large times. The authors in [41] proved uniform ultimate boundedness for a fairly general class of finite-dimensional nonlinear dynamic systems with continuous state feedback. In all references cited above, it can be stated that the so-called "matching assumptions" constitute sufficient conditions for a given uncertain system to be stabilizable. However, even for uncertain linear systems, these matching assumptions are known to be unduly restrictive. Indeed, it has been shown in [42,44] that there exist many uncertain linear systems which fail to satisfy matching conditions and yet are nevertheless stabilizable. Consequently recent research effort have been directed towards developing control schemes which will stabilize a larger class of systems than those which satisfy the matching conditions. related studies have been also reported in [45-51].

Numerous researchers have been conducted on controllers stabilizing systems with unmatched uncertainties [42,101-103]. It has been shown in [42,101] that the norm of the unmatched portion of the uncertain term must be smaller than a certain threshold value. The authors in [102] considered systems in which the uncertainty satisfies generalized matching conditions, i.e. structural conditions which are less restrictive than the matching conditions. There (as in the matched case) the norm bounds on the uncertain terms can be arbitrarily large. Linear time-invariant systems with scalar control input are treated in [103].

Other problems have been reported in [104-114]. References [104-106] deal with systems in which the uncertainty bounds are not known exactly but depend on unknown constants; the controllers presented there are parameter adaptive controllers. Problems in which one wishes to keep the system state within or outside a prescribed region of sate space have been considered in [106-110]. Related studies concerning systems with delay and discrete-time systems have been also reported in [111-115].

At a latter stage of development, the robustness of these controllers in the presence of singular perturbations have been considered in [116-119]. Moreover, the situation in which the full state of the system is not available for measurement has been treated in [82,120-125]. A nice review paper on recent trends in the control of uncertain dynamical systems is Corless and Leitmann [126].

It should be emphasized that controllers whose designs are based on Lyapunov theory have been applied to a variety of engineering problems including the tracking control of robotic manipulators [127-130], the control of structures in the presence of seismic excitations [131,132] and aerospace control problems [133-134]. Experimental results are contained in [128]. Applications to economic systems may be found in [135-137]. Harvesting problems are treated in [104,132] and river pollution control problems are considered in [138-142].

1.2 Organization of the Book

This book is devoted to the development of efficient control methodologies for systems whose dynamics include either unmodeled high-frequency parasitics or unknown but bounded parameters (henceforth termed partially-known dynamical systems). In order to achieve this objective, the book is organized as follows :

In Chapter 2, we present a new adaptive law for the robust adaptive control of plants with unmodeled high frequency dynamics. In the regulation case the adaptive system has bounded solutions. Stable performance is still guaranteed when the effect of high frequency parasitics is considered. It has been shown that the adaptive gain, the initial conditions, the mode-separation ratio and the magnitude and periodicity of the reference input sequence are important factors in the design of stable adaptive control schemes.

Some robustness properties of model reference adaptive schemes are analyzed in Chapter 3, for linear, discrete-time plants having unmodeled high frequency dynamics and/or external bounded disturbances. The stability behaviour with respect to reference model-plant order mismatch is examined. Three modified adaptation schemes are proposed. It is established that for bounded parasitics and input signals , reduced order controllers can be designed to yield uniformly asymptotically stable adaptive systems within a bounded region. Stable performance is still guaranteed when the effect of unmodeled parasitics is considered, but within a prescribed region of attraction.

Chapter 4 is concerned with the stabilization problem of a class of linear time - invariant , discrete systems with additive-type uncertainties. The dynamical system contains uncertain elements which are known to belong to prescribed compact bounding intervals. In addition, we consider the system to be corrupted by uncertain bounded inputs. A two-part switching feedback controller structure is developed in order to stabilize the uncertain dynamical system. The form of the controller (a linear part + a nonlinear part) guarantees global uniform ultimate boundedness under the validity of matching conditions. The performance of the uncertain system under the application of the proposed controller is compared with the application of purely linear one and gives superiority of using the former structure than the latter one.

Next, in Chapter 5, we have tried to relax the matching conditions. Instead, the stabilization problem is tackled via developing two-level control scheme. The asymptotic stability of the uncertain system is guaranteed and the control structure reliability is achieved. However, all the above feedback control system designs require the availability of the state of the controlled plant. Since the plant state variables are not generally available for direct measurement in practice, a means for estimating these variables is required. Attention is restricted to this requirement, and in addition, our analysis permits the controller to be nonlinear. Sufficient conditions in a form of inequalities are developed which indicate how large the excursions of the uncertain parameters can be. It has been shown that, for admissible uncertainties leading to the satisfaction of these inequalities, the controlled system guarantees asymptotic stability.

Chapter 6 is devoted to the development of new decentralized and hierarchical control techniques for linear interconnected , uncertain dynamical systems with additive-type bounded uncertainties. The overall system is decomposed into N-lower order subsystems, each containing uncertain elements and is corrupted by uncertain bounded disturbances. It has been shown that the proposed decentralized and hierarchical control structures guarantee global uniform ultimate boundedness behaviour for the decomposed subsystems. Sufficient conditions are given for the stability of the global system when driven with proposed control schemes and in the presence of interconnections as well as the bounded uncertainties in these interconnections. It has been established that with the satisfaction of these conditions and/or validity of the uncertainty matching structure, the developed decentralized and hierarchical control strategies provide robust design schemes ; that is , they are insensitive to the structural perturbations either between the subsystems and/or in the communication network between the two-level hierarchical structure. Furthermore, these design algorithms are insensitive to the parameter perturbations in their bounded ranges.

The same results are achieved in Chapter 7 for discrete time large-scale interconnected systems containing uncertain elements and subject to uncertain inputs.

Finally, the book is concluded in Chapter 8 and several open problems, for future research, are outlined.

It is worth mentioning that a more detailed description of our research work and its relationship to previous work in the area will be found at the beginning of each Chapter.

CHAPTER 2

CONTINUOUS SYSTEMS WITH REDUCED MODELS

2.1 Introduction

The problem of regulating a system with unknown parameters has been under investigation for quite long time [4,52]. Adaptive techniques provide an efficient method of handling plant uncertainty by adjusting the controller parameters on-line to optimize system performance. Global convergence has been established for a wide range of model reference adaptive control algorithms applicable to both continuous and discrete systems [13-17,52,54]. It has been most of the time assumed that the unknown plant and reference model structure are matched. Such an assumption is likely to be violated in applications, and thus attention is directed towards examining the robustness of adaptive schemes with respect to such modeling errors [32,34,55]. Recently, Ioannou and Kokotovic [34] restricted the input frequencies to so-called dominantly rich inputs, introduced a decay term in the adaptive law (σ-modification) and proved that these modifications guarantee the boundedness of all signals and their convergence to a residual set whose size depend on the disturbance and the mode-separation ratio. However,this procedure introduces a bias in the control parameter estimates so that they don't converge to the true values even when external disturbances are not present [31]. Moreover, pursting phenomena [56,57] can occur slowly due to the decaying term suggested in [32,34] resulting in sudden intermittent output error "bursts" followed by a long period of the apparent behaviour of the system or even complete instability of the system. Based on this, the occurrence of bursts in such systems contradicts one result reported in

[32,34] which states that the residual output error is small if σ is small. In fact, one can only guarantee that the "mean value" of the error is small if σ is small [57].

More recently, Narendra and Annaswamy [58] replaced the constant σ in [32] by a term proportional to $|e_1|$ where e_1 is the output error. This modification, referred to as $\sigma|e_1|$-modification is shown in [58] to improve the performance of plants with unknown parameters in the parasitic free case while retaining the advantage of assuring robustness in the presence of bounded disturbances, without requiring additional information about the plant or the disturbances.

The present work extends the idea of [58] and examines the problem of adaptively controlling reduced-order, continuous-time plants with unmodeled high frequency dynamics. The analysis is performed when the system under consideration operates in the regulation as well as tracking modes. We emphasize that the results of this Chapter can be summarized as follows : (1) the $\sigma|e_1|$-modification adaptation law can improve the dynamic performance of the closed-loop system in the presence of parasitics without requiring additional information about the parasitics; (2) an estimate for the value of σ is obtained in terms of the adaptation gain ; (3) in the regulation case, the bursts phenomena disappeared and the closed-loop system is asymptotically stable for unity adaptation gain ($\gamma=1$); (4) an estimate of a region of attraction from which all sequences converge to a bounded set about the equilibrium and finally (5) the adaptive gain, the initial conditions,the mode-separation ratio and the magnitude and periodicity of the reference input sequence are important factors in the design of stable adaptive control schemes.

2.2 Scalar Reduced-order Adaptive Control Problem

As a simple example consider a second-order plant

$$\dot{x}_p = a_p x_p + 2z - u \qquad (2.1)$$

$$\mu \dot{z} = -z + u \qquad (2.2)$$

in which the output x_p with unknown constant parameters a_p and μ, is required to track the state x_m of a first-order model :

$$\dot{x}_m = -a_m x_m + r \quad ; a_m > 0 \qquad (2.3)$$

where u(t) is the control input and r(.) is a bounded piecewise continuous functions. As in [32],the model-plant mismatch is due to some parasitic time constants which appear as multiples of a singular perturbation parameter μ and introduce the parasitic state η. In (2.1),(2.2) the parasitic state is defined as $\eta = z - u$ resulting in the following representation:

$$x_p = a_p x_p + 2 \quad + u \qquad (2.4)$$

$$\mu \dot{\eta} = - \eta - \mu \dot{u} \qquad (2.5)$$

where the dominant part (2.4) and parasitic part (2.5) of the plant appear explicitly [32].

The adaptive controller is chosen in the standard form as

$$u(t) = r(t) - \theta(t) x_p(t) \qquad (2.6)$$

where the control parameter $\theta(t)$ is adjusted using the available input-output data. From (2.4)-(2.6), the error and parasitic equations can be expressed as :

$$\dot{e}_1 = -a_m e_1 - \phi x_p + 2\eta$$

$$\mu\dot{\eta} = -\eta - \mu[\dot{r} - \dot{\theta} x_p - \theta \dot{x}_p]$$

where $\phi = \theta - \theta^*$, $\theta^* = a_p + a_m$, and $e_1 = x_p - x_m$. The adaptive law in the ideal case [13], modified in [32] and later modified in [58] are as follows :

$$\dot{\theta} = \dot{\phi} = \gamma e_1 x_p \quad ; \gamma > 0 \quad [13]$$

$$\dot{\theta} = \dot{\phi} = \gamma e_1 x_p - \sigma\theta \quad ; \sigma > 0, \gamma > 0 \quad [32]$$

$$\dot{\theta} = \dot{\phi} = \gamma e_1 x_p - \sigma|e_1|\theta \quad ; \sigma > 0, \gamma > 0 \quad [58]$$

In our work, we use the third form for the adaptation process and study the dynamic behaviour of the solutions of the resulting equations :

$$\dot{e}_1 = -a_m e_1 - \phi(e_1 + x_m) + 2\eta \tag{2.7}$$

$$\mu\dot{\eta} = -\eta + \mu[\gamma e_1(e_1 + x_m)^2 - \theta(\theta - a_p + \sigma|e_1|)(e_1 + x_m)$$
$$+ 2\theta\eta + \theta r - \dot{r}] \tag{2.8}$$

$$\dot{\theta} = \dot{\phi} = \gamma e_1(e_1 + x_m) - \sigma|e_1|\theta \quad ; \sigma > 0, \gamma > 0 \tag{2.9}$$

For the case without parasitics ($\mu = 0$) and unity adaptation gain ($\gamma = 1$), Narendra et al [58] proved that the $\sigma|e_1|$-modification scheme (2.9) can improve the performance of the system in all aspects while retaining the advantage of assuring robustness in the presence of bounded disturbances, without requiring additional information about plant or disturbances. In this respect, it is similar to the adaptive law suggested in [32]. However, in the ideal case, the $\sigma|e_1|$-scheme (2.9) results in exponential stability of the origin of the error equations if the reference input is persistently exciting with a large amplitude [58]. In the sequel, we test the adaptation law (2.9) for the system (2.7) and (2.8) in the regulation as well as tracking modes.

2.3 Adaptive Regulation

In the regulation problem [i.e. $r(t) = 0$, $x_m(t) = 0$, $e_1(t) = x_p(t)$], the system dynamic error equations (2.7)-(2.9) becomes :

$$\dot{x}_p = (a_p - \theta)\, x_p + 2\eta \tag{2.10}$$

$$\mu\dot{\eta} = -\eta + \mu\,[\gamma x_p^3 - \theta(\theta - a_p + \sigma|x_p|)\, x_p + 2\theta\eta] \tag{2.11}$$

$$\dot{\theta} = \gamma x_p^2 - \sigma|x_p|\theta \tag{2.12}$$

and the objective is to control the plant in such fashion that the dominant state x_p or equivalently the error e_1 goes to zero despite the presence of parasitics while assuring that all the signals in the closed-loop system (2.10)-(2.11) remain uniformly bounded. The following theorem establishes these stability properties for the system at hand.

Theorem 2.3.1

There exist positive scalars μ^*, t_1, σ, β, c_1-c_4 and $\alpha < 1/2$ such that every solution of (2.10)-(2.12) starting at $t = t_0$ from the set

$$\mathcal{R}_{s1}(\mu) = \{(x_p, \theta, \eta): \ |x_p| < c_1\,\mu^{-\alpha},\ |\theta| < c_2\,\mu^{-\alpha},\ |\eta| < c_3\,\mu^{-1/2-\alpha}\} \tag{2.13}$$

crosses the target set

$$\mathcal{R}_{t1}(\mu) = \{(x_p, \theta, \eta): \ [(\theta_1 - a_p)/4]\,|x_p|^2 + \beta|\theta - \theta_1|^2 +$$

$$+ 1/2 \; |\eta|^2 \le (c_4 \; \sigma/4\gamma)|\theta_1|^4 \; \} \qquad (2.14)$$

at $t=t_1$ and settles in $\mathcal{R}_{t1}(\mu)$ thereafter where

$$\sigma \le \gamma(\theta_1 - a_p), \; \beta \le \sigma \; c_1 \; \mu^{-\alpha} \; / \; (2\gamma) \qquad (2.15)$$

and θ_1 is a finite constant $> a_p$.

<u>Proof</u>

Consider the Δ-parameterized function

$$V(x_p, \theta, \eta, \Delta) = 1/2 \; x_p^2 + (1/2\gamma)(\theta - \theta_1)^2 + (\mu/2)(\eta + \Delta)^2 \qquad (2.16)$$

Following [32], we can see that for μ, $c_0 > 0$, $\alpha < 1/2$ and $\Delta = 2\,x_p$, in order to facilitate the analysis, the equality $V = c_0\;\mu^{-2\alpha}$ characterizes a closed surface $\mathcal{S}(c_0, \mu, \alpha)$ in the composite space R^3. The time derivative of (2.16) along the solution of (2.10)-(2.12) is

$$\dot{V}(x_p, \theta, \eta) = -(\theta_1 - a_p)\; x_p^2 - (\sigma/\gamma)\; |x_p| \; \theta \; (\theta - \theta_1) - \eta^2 +$$

$$+ \mu \; (\eta + 2x_p) \; \{ \gamma \, x_p^3 - \theta(\theta - a_p + \sigma |x_p|)x_p + 2\theta\,\eta + 2a_p x_p -$$

$$- 2\theta x_p + 4\eta \; \} \qquad (2.17)$$

For all sequences $\{x_p(t)\}$, $\{\theta(t)\}$, $\{\eta(t)\}$ originated in $\mathcal{R}_{s1}(\mu)$ we can express (2.17), after completing squares and grouping terms, as :

$$\dot{V}(x_p, \theta, \eta) \le -|x_p|^2 \; \{ \; (\theta_1 - a_p)/2 - \mu[2\gamma \, x_p^2 - 2\theta(\theta - a_p + \sigma|x_p|) +$$

$$+ 4a_p - 4\theta] - \mu^2 \; [\gamma \, x_p^2 - \theta(\theta - a_p + \sigma|x_p|) + 2a_p + 2\theta + 8]^2\} -$$

$$- 1/4 \{ \eta - \mu x_p [2 \gamma x_p^2 - 2\theta(\theta - a_p + \sigma|x_p|) + 4a_p + 4\theta + 16]\}$$

$$- (\theta-\theta_1)^2.\{(\sigma/2\gamma)|x_p| - \beta\} - \eta^2\{1/4 - \mu (2\theta+4)\} -$$

$$-(\sigma/2\gamma)|x_p|.|\theta|^2 - |x_p|^2.\{(\theta_1-a_p)/4 - (\sigma/4\gamma)\} -$$

$$-[(\theta_1-a_p)/4]|x_p|^2 - \beta|\theta-\theta_1|^2 - 1/2|\eta|^2 + (\sigma/4\gamma)|\theta_1|^4 \qquad (2.18)$$

For x_p, θ, η inside $\mathcal{S}(c_0,\mu,\alpha)$, (2.18) is simplified to

$$\dot{V}(x_p,\theta,\eta) \le -|x_p|^2 \{ (\theta_1-a_p)/2 - \beta_1 \mu^{1-2\alpha} - \beta_2 \mu^{2(1-2\alpha)} \} -$$

$$-|\theta-\theta_1|^2.\{(\sigma c_1/2\gamma) \mu^{-\alpha} - \beta\} - |\eta|^2 \{1/4 - 2 c_2 \mu^{1-\alpha} -$$

$$- 4\mu \} - |x_p|^2.[(\theta_1-a_p)/4 - (\sigma/4\gamma)] - [(\theta_1-a_p)/4]|x_p|^2 -$$

$$- \beta|\theta-\theta_1|^2 - 1/2|\eta|^2 + (\sigma/4\gamma)|\theta_1|^4 \qquad (2.19)$$

On choosing $\sigma \le \gamma(\theta_1-a_p)$ and $\beta \le (\sigma c_1/2\gamma)\ \mu^{-\alpha}$ (2.20)

there exist a μ^* such that for each $\mu \in (0, \mu^*]$

$$(\theta_1-a_p)/2 \ge \mu^{1-2\alpha} (\beta_1 + \beta_2 \mu^{1-2\alpha}) \quad \text{and} \quad 1/4 \ge 2c_2 \mu^{1-\alpha} + 4\mu \qquad (2.21)$$

$$\therefore \dot{V}(x_p,\theta,\eta) < -[(\theta_1-a_p)/4]|x_p|^2 - \beta|\theta-\theta_1|^2 - 1/2|\eta|^2 +$$

$$+ (\sigma/4\gamma)|\theta_1|^4 \qquad (2.22)$$

for all $\mu \in (0,\mu^*]$ and x_p, θ, η inside $\mathcal{S}(c_0,\mu,\alpha)$.
Noting that $\mathcal{R}_{t1}(\mu)$ is uniformly bounded region, it is readily evident that there exist constants c_1 to c_3 such that the solutions of (2.10)-(2.12) starting from $\mathcal{R}_{s1}(\mu)$ enter the target set $\mathcal{R}_{t1}(\mu)$ and such that $\mathcal{R}_{t1}(\mu) \subset \mathcal{R}_{s1}(\mu) \subset \mathcal{S}(c_0,\mu,\alpha)$ as shown in Fig. 2.1. Observe that $\dot{V}(x_p,\theta,\eta) < 0$

everywhere in $\mathcal{S}(c_0,\mu,\alpha)$ except possibly in $\mathcal{R}_{t1}(\mu)$, and $V(x_p,\theta,\eta)$ is monotonically non-increasing in $\mathcal{R}_{t1}(\mu)/\mathcal{R}_{s1}(\mu)$. This means that there exist constants $t_1 \geq t_0$, $c_4 \geq 1$ such that any solution starting at $t=t_0$ from $\mathcal{R}_{s1}(\mu)$ will cross the target $\mathcal{R}_{t1}(\mu)$ at $t=t_1$ and settles in $\mathcal{R}_{t1}(\mu)$ for all $t>t_1$. Therefore, $\lim_{t\to\infty} \dot{V}(x_p(t),\theta(t),\eta(t)) = 0$, i.e. $x_p(t) \longrightarrow 0$, $\eta(t) \longrightarrow 0$ and $\theta(t) \longrightarrow$ constant as $t \longrightarrow \infty$ which completes the proof. ▪▪▪

As an illustration of the stability properties established by Theorem 2.3.1, simulation results for (2.10)-(2.12) with $a_p=4$, $\theta_1=7$ and different values of μ, σ and initial conditions are plotted in Figs (2.2-2.4). In all cases, it is shown that the system is asymptotically stable ($x_p \longrightarrow 0$ and V is bounded). It is worth mentioning that, we took the adaptive gain $\gamma=1$. For large values of the adaptive gain (we try the value $\gamma=30$), the error equations (2.10)-(2.12) give unbounded solution. For the purpose of comparison with the σ-modification adaptation law [32], increasing the values of $x_p(0)$ from 1 to 2.4 and/or μ from 0.05 to 0.07 doesn't affect the stability properties as shown in Figs. 2.2 and 2.3.

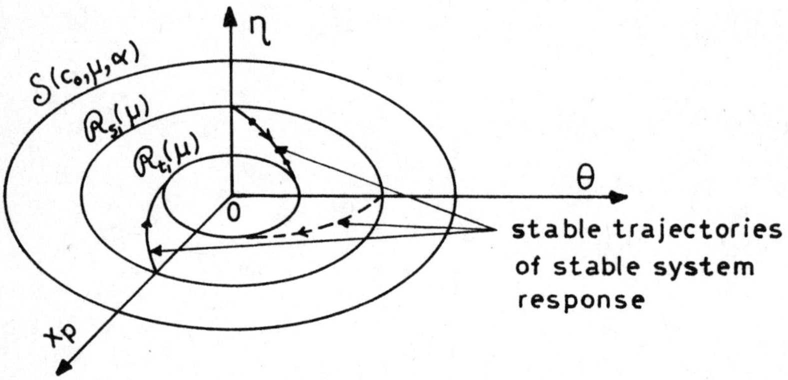

Figure 2.1: Geometry associated with Lyapunov criteria.

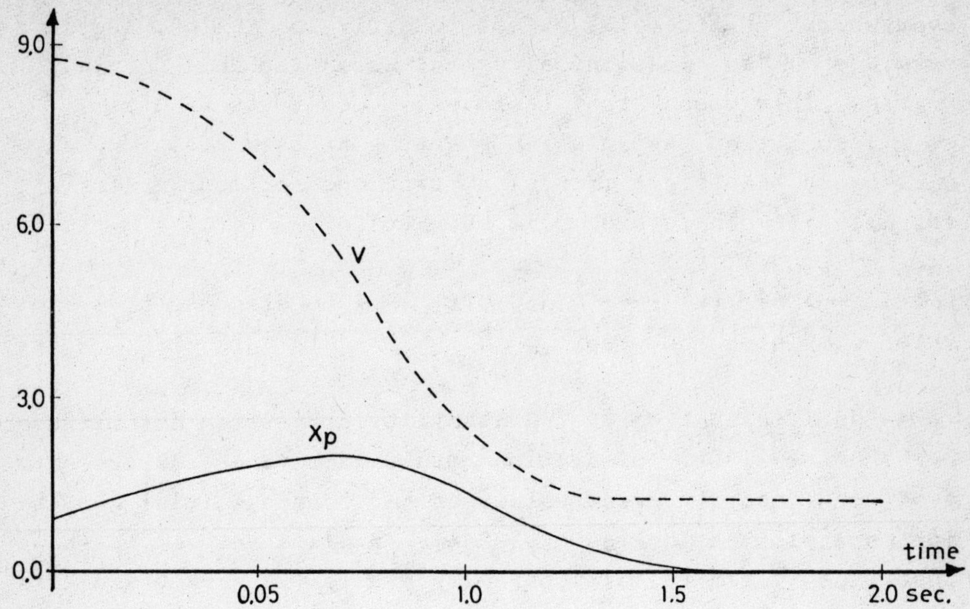

Fig. 2.2: Adaptive regulation for μ=0.05, σ=0.1, γ=1, $x_p(0)$=1, $\eta(0)$=1 and $\theta(0)$=3.

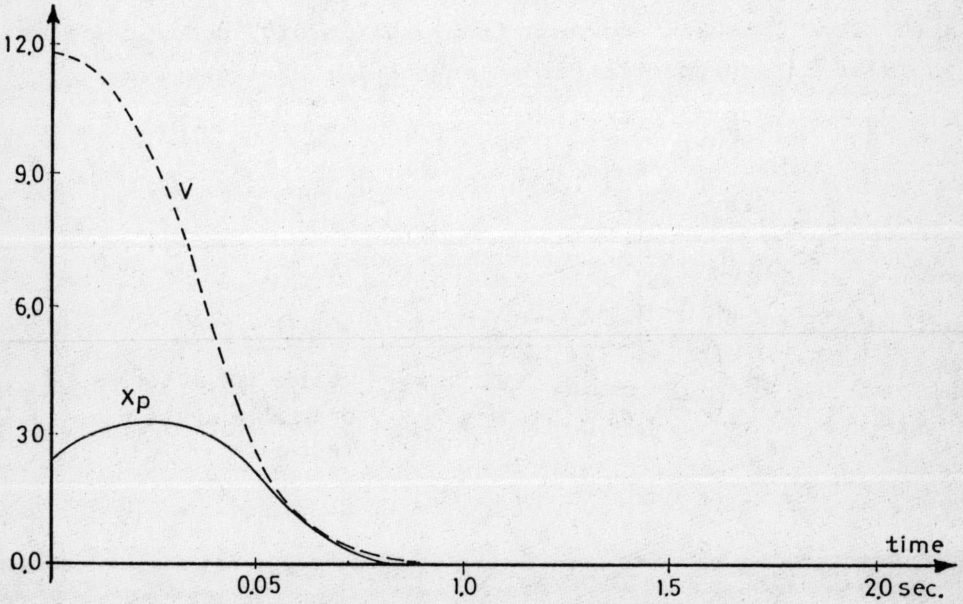

Fig. 2.3: Adaptive regulation for μ=0.05, σ=0.1, γ=1, $x_p(0)$=2.4, $\eta(0)$=1 and $\theta(0)$=3.

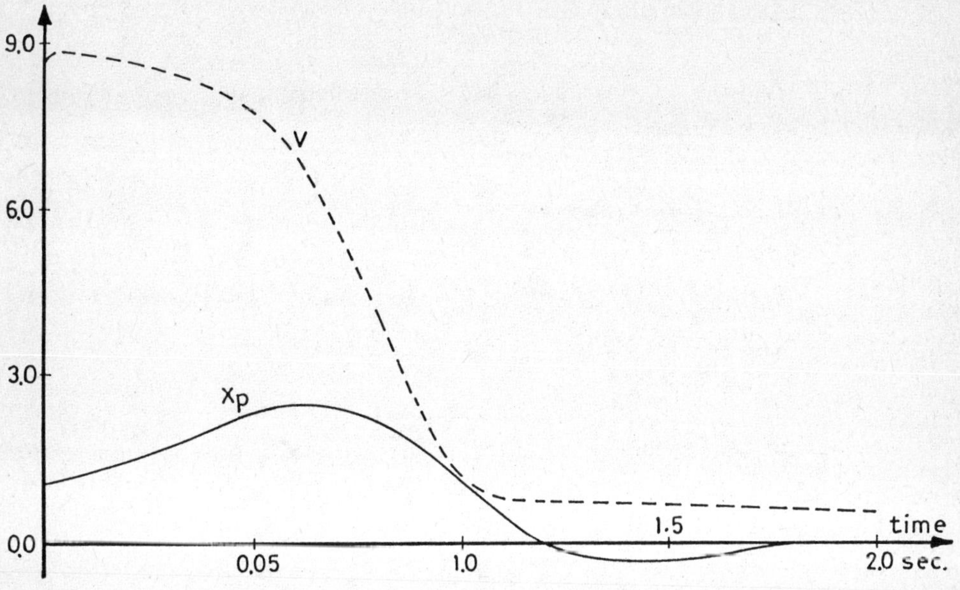

Fig. 2.4: Adaptive regulation for μ=0.07, σ=0.1, γ=1, $x_p(0)$=1, $\eta(0)$=1 and $\theta(0)$=3.

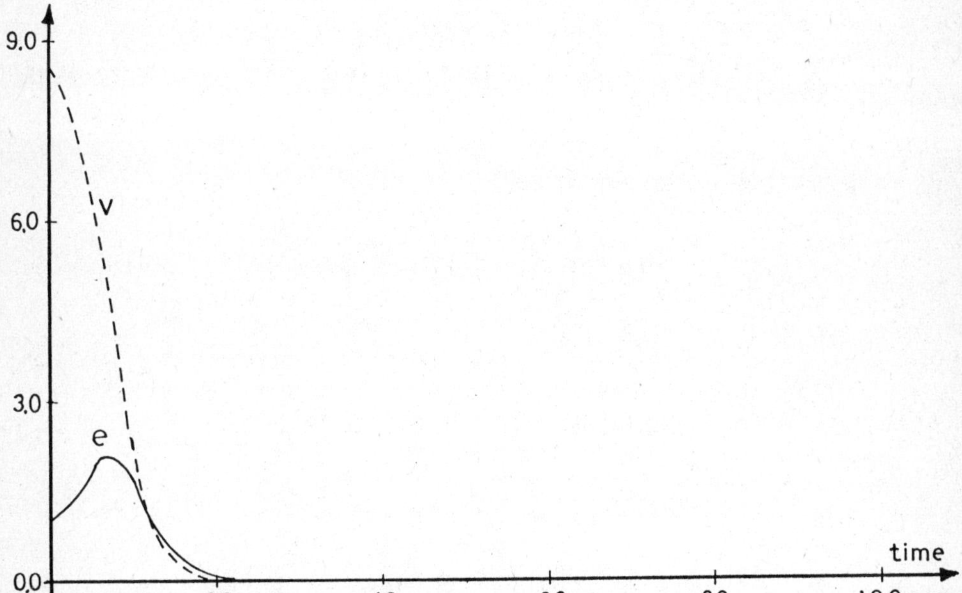

Fig. 2.5: Adaptive tracking for μ=0.01, σ=0.1, γ=1, e(0)=1, $\eta(0)$=1, $\theta(0)$=3 and r(t)=3 sin 2t.

2.4 Adaptive Tracking

In this case, we will study the stability properties for the following dynamic error equations :

$$\dot{e}_1 = -a_m e_1 - \phi(e_1 + x_m) + 2\eta \tag{2.23}$$

$$\mu\dot{\eta} = -\eta + \mu\,[\gamma e_1(e_1 + x_m)^2 - \theta(\theta - a_p + \sigma|e_1|)(e_1 + x_m) + 2\theta\eta + \theta r - \dot{r}] \tag{2.24}$$

$$\dot{\theta} = \dot{\phi} = \gamma e_1(e_1 + x_m) - \sigma|e_1|\theta \quad ; \sigma > 0,\ \gamma > 0 \tag{2.25}$$

Theorem 2.4.1

Let the reference input $r(t)$ satisfy :

$$|r(t)| < r_1 \quad , \quad |\dot{r}(t)| < r_2 \qquad \forall\ t > t_0 \tag{2.26}$$

where r_1, r_2 are given positive constants. Then there exist positive scalars μ^*, t_1, σ, β, $c_1 - c_4$ and $\alpha<1/2$ such that every solution of (2.23)-(2.25) starting at $t = t_0$ from the set :

$$\mathcal{R}_{s2}(\mu) = \{(e,\theta,\eta): |e| < c_1\mu^{-\alpha},\ |\theta| < c_2\mu^{-\alpha},\ |\eta| < c_3\mu^{-1/2-\alpha}\} \tag{2.27}$$

crosses the target set

$$\mathcal{R}_{t2}(\mu) = \{(e,\theta,\eta): a_m/8\,|e|^2 + \beta|\theta-\theta^*|^2 + 1/8\,|\eta|^2 < c_4[(\sigma/4\gamma)|\theta^*|^4 + \mu^2(1+4/a_m)|2\,\theta^* x_m - \dot{r}|^2]\} \tag{2.28}$$

at $t=t_1$ and settles in $\mathcal{R}_{t2}(\mu)$ thereafter where

$$(2\gamma\Phi_3/c_1)\mu^{2-\alpha} \le \sigma \le \gamma a_m/2 \quad , \beta \le (\sigma c_1/2\gamma)\mu^{-\alpha} \qquad (2.29)$$

and Φ_3 is some positive constant.

Proof

Consider the function

$$V(e,\theta,\eta) = 1/2\, e^2 + (1/2\gamma)(\theta-\theta^*)^2 + (\mu/2)(\eta + 2e)^2 \qquad (2.30)$$

Similarly, as shown in [32] and the proof of Theorem 2.3.1, for μ, $c_0 > 0$ and $\alpha < 1/2$ the equality $V = c_0\,\mu^{-2\alpha}$ characterizes a closed surface $\mathcal{S}(c_0,\mu,\alpha)$ in the composite space R^3. The time derivative of $V(e,\theta,\eta)$ along the trajectories (2.23)-(25) is

$$\dot{V}(e,\theta,\eta) = -a_m\, e^2 - \sigma|e|\,\phi\theta/\gamma - \eta^2 + \mu\,(\eta + 2e)\,\{\gamma e^3 +$$

$$+ 2\gamma e^2\, x_m + \gamma e\, x_m^2 - \theta(\theta - a_p + \sigma|e|)\,(e + x_m) + 2\theta\eta + \theta\, r$$

$$- \dot{r} - 2\, a_m\, e - 2\,\phi\,(e + x_m) + 4\eta\} \qquad (2.31)$$

In view of (2.27), (2.31) can be written as :

$$\dot{V}(e,\theta,\eta) \le -|e|^2\,\{a_m/4 - \mu[2\gamma e^2 + 4\gamma e x_m + 2\gamma x_m^2 - 2\theta\,(\theta -$$

$$- a_p + \sigma|e|) - 4a_m - 4\phi] - \mu^2\,[\gamma e^2 + 2\gamma e x_m + \gamma x_m^2 - \theta\,(\theta -$$

$$- a_p + \sigma|e|) - 2a_m - 2\phi + 4\theta + 8]^2\} - 1/4\,\{\eta - 2\,\mu e\,[\gamma e^2 +$$

$$+ 2\gamma e x_m - \theta\,(\theta - a_p + \sigma|e|) - 2a_m - 2\phi + 4\theta + 8]\}^2 -$$

$$- 1/4\,[\eta - 2\,\mu(2\theta^* x_m - \dot{r})]^2 - 1/4\,\{\eta - 2\,\mu\theta\,[r - 2x_m - (\theta -$$

$$- a_p + \sigma|e|)\, x_m\}^2 - |\eta|^2\,\{1/8 - \mu\,(2\theta + 4)\} - a_m/4\,\{e -$$

$$- 4(\mu\theta/a_m)\,[r - 2x_m - (\theta - a_p + \sigma|e|)\,x_m]\}^2 + \mu^2\,(1 + 4/a_m)\bullet$$

$$\bullet\,|2\theta^* x_m - \dot{r}|^2 - (a_m/4)\,\{e - 4\,(\mu/a_m)\,|2\theta^* x_m - \dot{r}|\}^2 -$$

$$- \theta^2\,\{(\sigma/2\gamma)|e| - 4(\mu^2/a_m)[r - 2x_m + (\theta - a_p + \sigma|e|)\,x_m]^2\}$$

$$- (1/8\gamma)|e|^2\,(\gamma a_m - 2\sigma) - (1/2\gamma)|\phi|^2\,(\sigma|e| - 2\gamma\beta) -$$

$$- (a_m/8)\,|e|^2 - \beta|\phi|^2 - 1/8\,|\eta|^2 + (\sigma/4\gamma)|\theta^*|^4 \qquad (2.32)$$

For all sequences $\{e(t)\},\{\phi(t)\},\{\eta(t)\}$ inside $\mathcal{S}(c_0,\mu,\alpha)$, (2.32) is simplified to

$$\dot{V}(e,\theta,\eta) \le -|e|^2\,\{a_m/4 - \Phi_1\,\mu^{1-2\alpha} - \Phi_2\,\mu^{2(1-2\alpha)}\} - |\eta|^2\bullet$$

$$\bullet\{1/8 - 2c_2\,\mu^{2-\alpha} - 4\mu\} + \mu^2\,(1 + 4/a_m)\,|2\theta^* x_m - \dot{r}|^2 -$$

$$-|\theta|^2\,\{(\sigma c_1/2\gamma)\,\mu^{-\alpha} - \Phi_3\,\mu^{2(1-\alpha)}\} - (1/8\gamma)|e|^2(\gamma a_m -$$

$$-2\sigma) - (1/2\gamma)|\phi|^2\,(\sigma c_1\,\mu^{-\alpha} - 2\gamma\beta) - (a_m/8)|e|^2 - \beta|\phi|^2$$

$$- (1/8)\,|\eta|^2 + (\sigma/4\gamma)|\theta^*|^4 \qquad (2.33)$$

for some positive constant Φ_1, Φ_2 and Φ_3. Now choose

$$(2\gamma\Phi_3/c_1)\,\mu^{2-\alpha} \le \sigma \le \gamma a_m/2\ ,\ \beta \le (\sigma c_1/2\gamma)\,\mu^{-\alpha} \qquad (2.34)$$

and $\alpha<1/2$. Then there exists a μ^* such that for each $\mu\in(0,\mu^*]$

$$a_m/4 \ge \mu^{1-2\alpha}\,(\Phi_1 + \Phi_2\,\mu^{1-2\alpha}) \quad \text{and} \quad 1/8 \ge 2c_2\,\mu^{1-2\alpha} + 4\mu \qquad (2.35)$$

Hence, for each $\mu \in (0,\mu^*]$ and all e,θ,η inside $\mathcal{S}(c_0,\mu,\alpha)$,

$$\dot{V}(e,\theta,\eta) < -(a_m/8)|e|^2 - \beta|\phi|^2 - (1/8)|\eta|^2 + (\sigma/4\gamma)|\theta^*|^4$$

$$+ \mu^2 (1+4/a_m)|2\theta^* x_m - \dot{r}|^2 \tag{2.36}$$

Due to the uniform boundedness of the input sequence r and $\dot{r}$, the target set $\mathcal{R}_{t2}(\mu)$ is uniformly bounded. It is thus clear that there exist constants $c_1 - c_3$ such that the solutions of (2.23)-(2.25) starting from $\mathcal{R}_{s2}(\mu)$ enter the set $\mathcal{R}_{t2}(\mu)$ and such that $\mathcal{R}_{t2}(\mu) \subset \mathcal{R}_{s2}(\mu) \subset \mathcal{S}(c_0,\mu,\alpha)$. We note that $\dot{V}(e,\theta,\eta) < 0$ everywhere in $\mathcal{S}(c_0,\mu,\alpha)$, except possibly in $\mathcal{R}_{t2}(\mu)$, and $V(e,\theta,\eta)$ is monotonically non-increasing in $\mathcal{R}_{t2}(\mu)/\mathcal{R}_{s2}(\mu)$. Consequently, there exist constants $t \geq t_0$ and $c_4 \geq 1$ such that any solution originating from $\mathcal{R}_{s2}(\mu)$ at $t = t_0$ will enter $\mathcal{R}_{t2}(\mu)$ at $t = t_1$ and resides there for $t \geq t_1$.

As an illustration, simulation results are summarized in Fig. 2.5-2.14 with $a_p=4$, $a_m=3$ and $\gamma=1$. In Fig. 2.5, the output error and the function $V(e,\theta,\eta)$, given by (2.16), are plotted for $\mu=0.01$, $\sigma=0.1$, $e(0)=1$, $\eta(0)=1$, $\theta(0)=3$ and $r(t)=3 \sin 2t$. It can be shown that the system behaves asymptotically stable even if μ is increased to 0.05 as shown in Fig. 2.6. However, increasing μ to 0.08 gives unbounded solutions for $\sigma\geq 0$. Keeping the same conditions as in Fig. 2.5 and increasing the value of e(0) to 2.5, we can achieve asymptotically stable system for $\sigma=0,0.1$ as shown in Figs. 2.7, 2.8, respectively. Figs. 2.9-2.11 show the effect of high frequency input ($r(t)=3 \sin 10t$) and the value of σ takes the values 0, 0.1 and 1 respectively. It can be observed that smaller values for σ gives more bounded behaviour. Also, bounded solutions can be obtained for large amplitude of the reference input as shown in Figs. 2.12 and 2.13 ($r(t)=15 \sin 2t$). In Fig. 2.14, we show the loss of exact convergence of the output error when $\mu=0$ due to the design parameter σ.

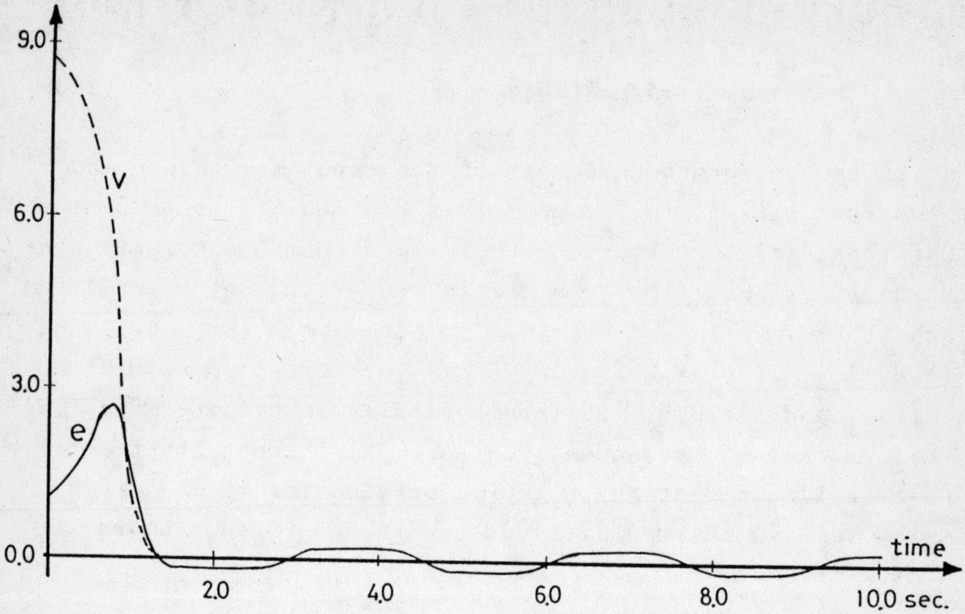

Fig. 2.6: Adaptive tracking for μ=0.05, σ=0.1, γ=1, e(0)=1, η(0)=1, θ(0)=3 and r(t)=3 sin 2t.

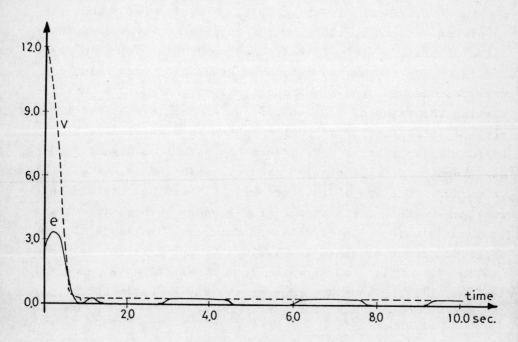

Fig. 2.7: Adaptive tracking for μ=0.05, σ=0.1, γ=1, e(0)=2.5, η(0)=1, θ(0)=3 and r(t)=3 sin 2t.

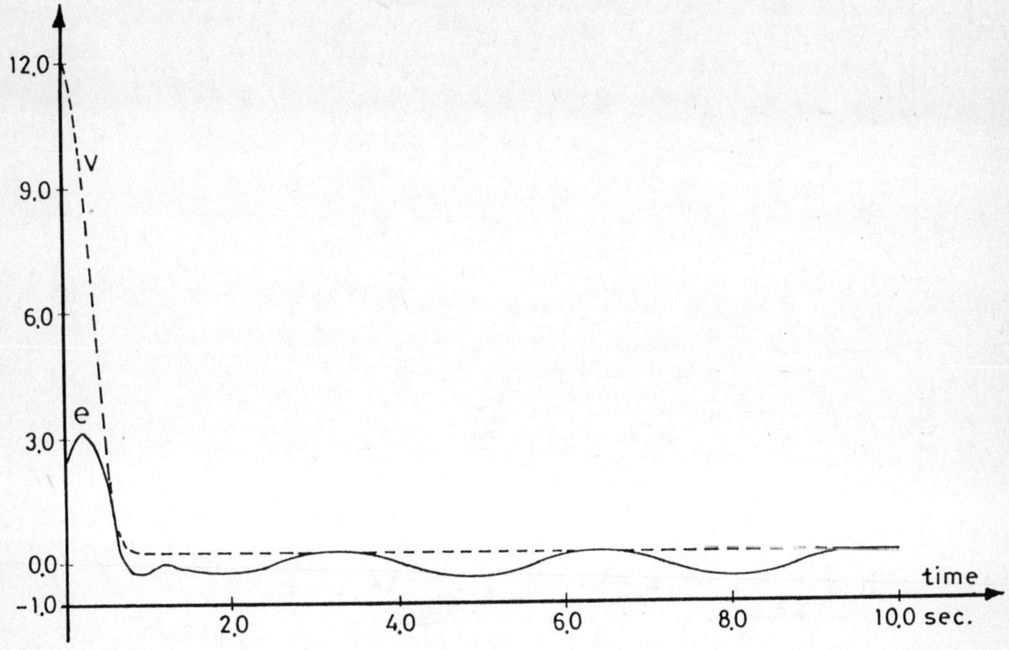

Fig. 2.8: Adaptive tracking for μ=0.05, σ=0, γ=1, e(0)=2.5, η(0)=1, θ(0)=3 and r(t)=3 sin 2t.

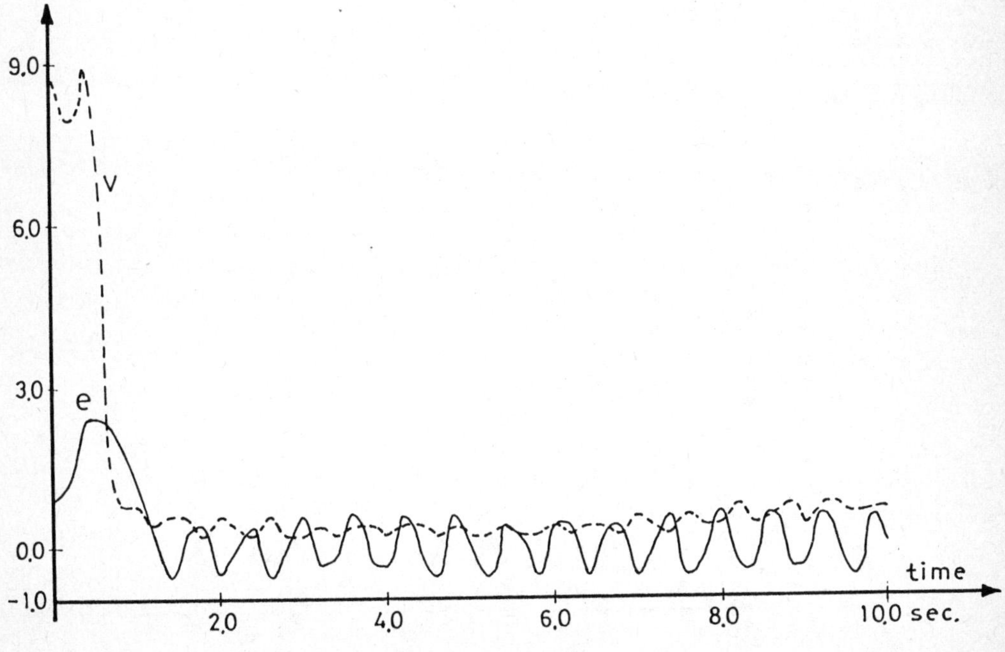

Fig. 2.9: Adaptive tracking for μ=0.05, σ=0, γ=1, e(0)=1, η(0)=1, θ(0)=3 and r(t)=3 sin 10t.

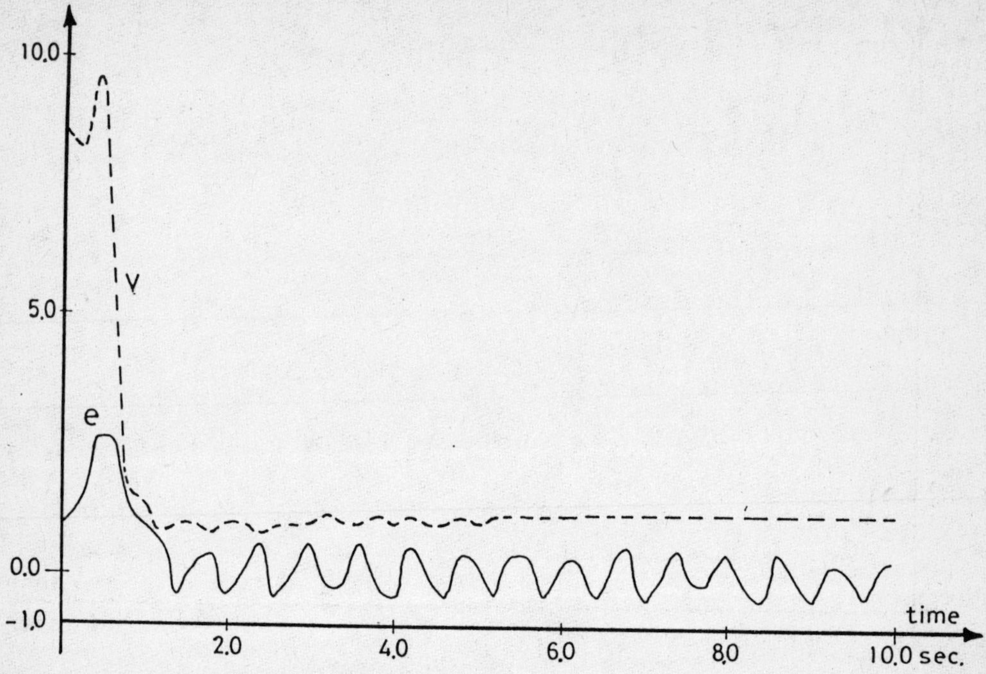

Fig. 2.10: Adaptive tracking for $\mu=0.05$, $\sigma=0.1$, $\gamma=1$, $e(0)=1$, $\eta(0)=1$, $\theta(0)=3$ and $r(t)=3 \sin 10t$.

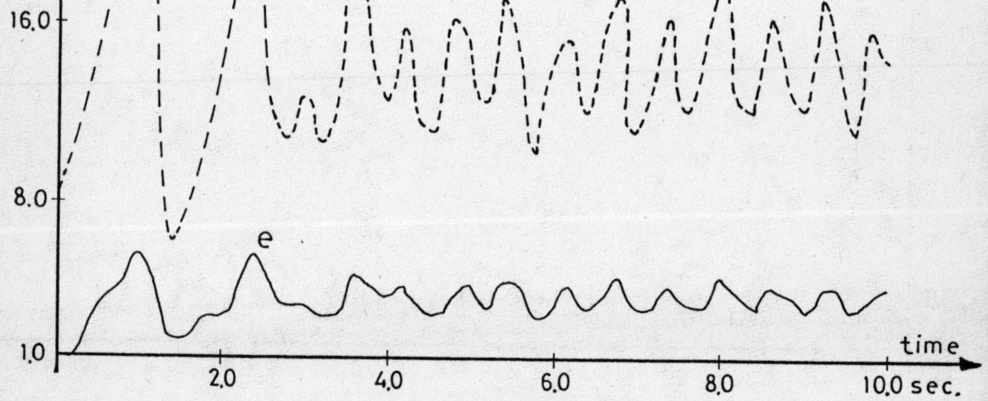

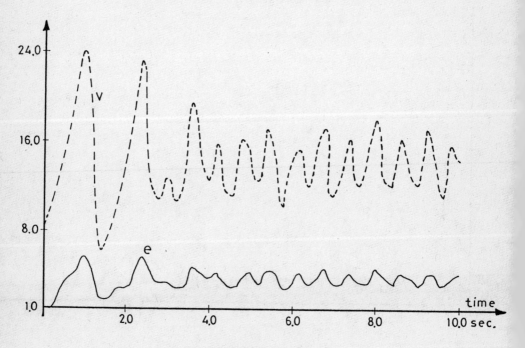

Fig. 2.11: Adaptive tracking for $\mu=0.05$, $\sigma=1$, $\gamma=1$, $e(0)=1$, $\eta(0)=1$, $\theta(0)=3$ and $r(t)=3 \sin 10t$.

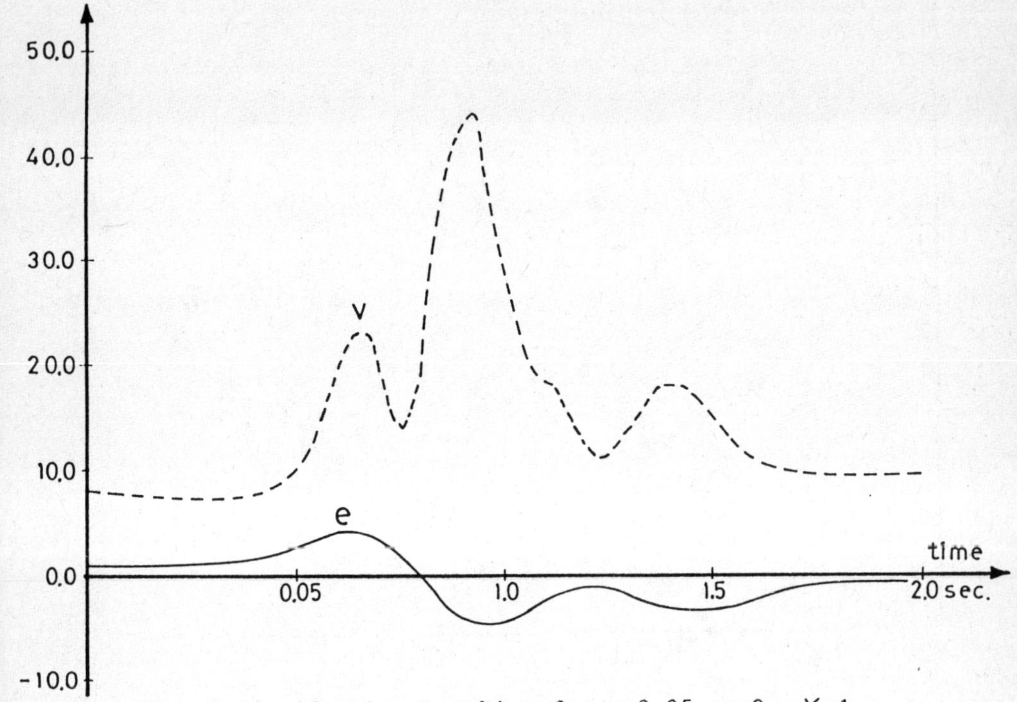

Fig. 2.12: Adaptive tracking for μ=0.05, σ=0, γ=1, e(0)=1, η(0)=1, θ(0)=3 and r(t)=15 sin 2t.

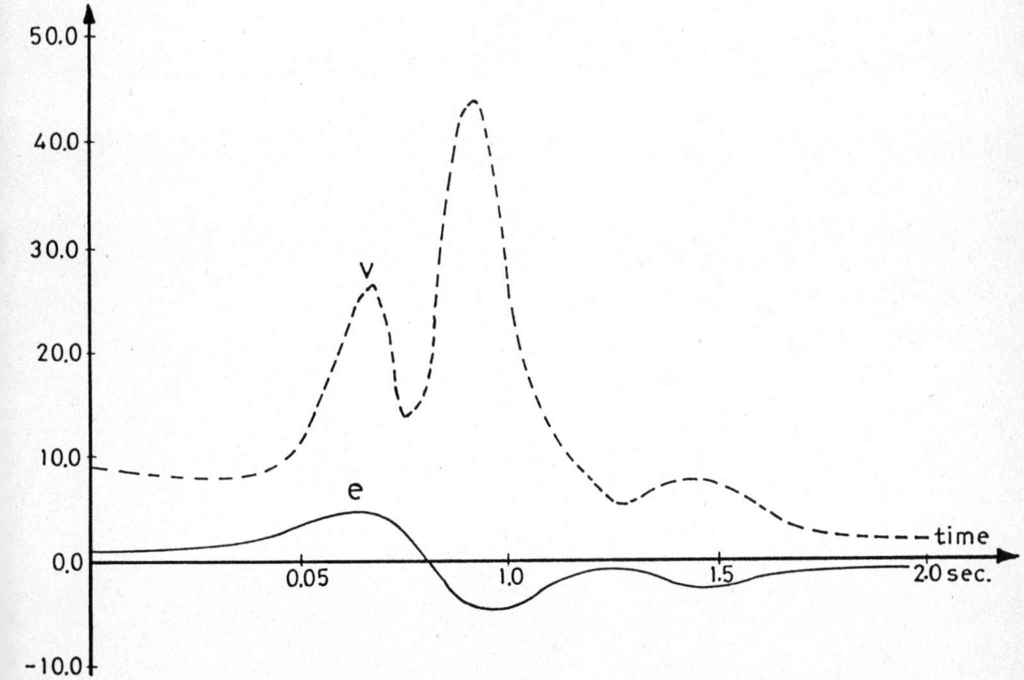

Fig. 2.13: Adaptive tracking for μ=0.05, σ=0.1, γ=1, e(0)=1, η(0)=1, θ(0)=3 and r(t)=15 sin 2t.

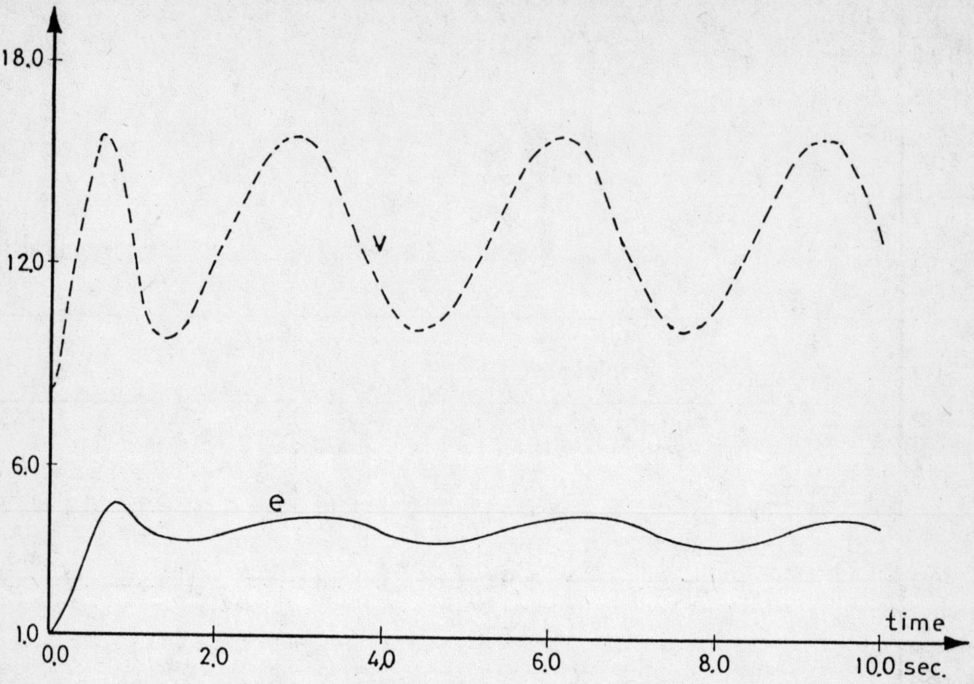

Fig. 2.14: Adaptive tracking for $\mu=0$, $\sigma=1$, $\gamma=1$, $e(0)=1$, $\eta(0)=1$, $\theta(0)=3$ and $r(t)=3 \sin 2t$.

2.5 Adaptive Control For SISO System With Parasitics.

Motivated by the above discussions and inspired by the pioneering results obtained by Ioannou and Kokotovic [32], we generalize the foregoing analysis to admit SISO plants in the presence of parasitics. A common model of a SISO plant with stable high frequency parasitics [32] is the so-called standard singular perturbation model

$$\dot{\underline{X}}_p = A_{11} \underline{X}_p + A_{12} \underline{Z} + \underline{b}_1 u \tag{2.35}$$

$$\mu \dot{\underline{Z}} = A_{21} \underline{X}_p + A_{22} \underline{Z} + \underline{b}_2 u \quad , \quad \text{Re } \lambda(A_{22}) < 0 \tag{2.36}$$

$$y_p = \underline{C}_0^T \underline{X}_p \tag{2.37}$$

where μ is a small positive parameter, $\underline{X}_p$ and $\underline{Z}$ are the n-dimensional dominant and the m-dimensional parasitic states, respectively, and u, y_p are the scalar input and output of the plant, respectively. The restriction of the output to be of the type (2.37) allows the fast parasitics to be weakly observable; that is observable through the slow part of the plant [32,34,55]. State Z is formed of a "fast transient" and a "quasi-steady state" defined as the solution of (2.36) with $\mu = 0$ [32,34]. This motivates the definition of the fast parasitic state as :

$$\underline{\eta} = \underline{Z} + A_{22}^{-1} (A_{21} \underline{X}_p + \underline{b}_2 u) \tag{2.38}$$

The substitution of (2.38) into (2.35)-(2.37) yields :

$$\dot{\underline{X}}_p = A_0 \underline{X}_p + \underline{b}_0 u + A_{12} \underline{\eta} \tag{2.39}$$

$$\mu \dot{\underline{\eta}} = A_{22} \underline{\eta} + \mu (A_1 \underline{X}_p + \underline{A}_2 u + A_3 \underline{\eta} + \underline{A}_4 \dot{u}) \tag{2.40}$$

$$y_p = \underline{C}_0^T \underline{X}_p \tag{2.41}$$

where

$$A_0 = A_{11} - A_{12} A_{22}^{-1} A_{21} \quad , \quad \underline{b}_0 = \underline{b}_1 - A_{12} A_{22}^{-1} \underline{b}_2 \ ,$$

$$A_1 = A_{22}^{-1} A_{21} A_0 \quad , \quad \underline{A}_2 = A_{22}^{-1} A_{21} \underline{b}_0 \ , \qquad (2.42)$$

$$A_3 = A_{22}^{-1} A_{21} A_{12} \quad , \quad \underline{A}_4 = A_{22}^{-1} \underline{b}_2$$

It can be emphasized that representation (2.39)-(2.42) is convenient since the dominant and parasitic parts appear explicitly. The following standard assumptions are made throughout the Chapter :

(i) $\lambda(A_4) < 0$,
(ii) the triple $\{A_0, \underline{b}_0, \underline{C}_0^T\}$ is completely controllable and completely observable.

Now, we consider the reduced-order adaptive control problem of the system (2.39)-(2.42) in which the output y_p is requited to track the output y_m of an nth -order reference model :

$$\dot{\underline{X}}_m = A_m \underline{X}_m + \underline{b}_m r \qquad (2.43)$$

$$y_m = \underline{C}_m^T \underline{X}_m \qquad (2.44)$$

whose transfer function $W_m(s)$ is given by

$$W_m(s) = \underline{C}_m^T (sI-A_m)^{-1} \underline{b}_m = K_m N_m(s)/D_m(s) \qquad (2.45)$$

is chosen to be strictly positive real and r(t) is a uniformly bounded reference signal; $|r(t)| \leq r_1$ and $|\dot{r}(t)| \leq r_2 \quad \forall \ t$. The reduced-order plant obtained by setting $\mu = 0$ in (2.39)-(2.41) has a transfer function

$$W_0(s) = \underline{C}_0^T (sI-A_0)^{-1} \underline{b}_0 = K_0 N_0(s)/D_0(s) \qquad (2.46)$$

which is assumed to be strictly positive real. The controller structure has the same form as [15] for the parasitic free plant ($\mu = 0$). In this structure, the input u(t) and output $y_p(t)$ are proceed to generate two (n-1)-dimensional auxiliary vectors $\underline{f}(t)$ and $\underline{h}(t)$ in the form :

$$\dot{\underline{f}}(t) = G\underline{f}(t) + \underline{g}u(t) \quad ; \quad \tilde{v}(t) = \underline{C}^T(t)\underline{f}(t) \tag{2.47}$$

$$\dot{\underline{h}}(t) = G\underline{h}(t)+\underline{g}y_p(t) \quad ; \quad \tilde{w}(t) = d_o(t)y_p(t)+\underline{d}^T(t)\underline{h}(t) \tag{2.48}$$

where G is an (n-1)x(n-1) stable matrix and $(G,\underline{g})$ is a controllable pair. The input of the plant is given by :

$$\begin{aligned} u(t) &= r(t) + \underline{\theta}^T(t)\, \underline{w}(t) \\ &= r(t) + \underline{\phi}^T(t)\, \underline{w}(t) + \underline{\theta}^*\, \underline{w}(t) \end{aligned} \tag{2.49}$$

where $\underline{\theta}^T(t) = [\underline{C}^T(t), d_o(t), \underline{d}^T(t)]$ is a (2n-1) vector of adjustable parameters, and $\underline{w}^T(t) = [\underline{f}^T(t), y_p(t), \underline{h}^T(t)]$ is an augmented state vector of dimension (2n-1). It has been shown in [15] that a constant vector $\underline{\theta}^*$ exists such that for $\underline{\theta}(t) = \underline{\theta}^*$ the transfer function of the parasitic-free plant (2.46) with the controller (2.47)-(2.49) matches that of the reference model (2.45).

Introducing $\underline{q}^T(t) = [X_p^T, \underline{f}^T, \underline{h}^T]$, $\tilde{A}_{12} = [A_{12}^T \;\; 0 \;\; 0]^T$, $\tilde{A}_1 = [A_1^T \;\; 0 \;\; 0]^T$, $\underline{b}_g^T = [\underline{b}_o^T, \underline{g}^T, 0]$ and

$$A_g = \begin{bmatrix} A_o + \underline{b}_o d^* \underline{C}_o^T & \underline{b}_o \underline{C}^{*T} & \underline{b}_o \underline{d}^{*T} \\ \underline{g} d^* \underline{C}_o^T & G + \underline{g}\underline{C}^{*T} & \underline{g}\underline{d}^{*T} \\ \underline{g}\underline{C}_o^T & 0 & G \end{bmatrix} \tag{2.50}$$

We now apply the controller (2.47)-(2.49) to the full-plant (2.39)-(2.41) to obtain the augmented feedback system :

$$\dot{\underline{q}} = A_g \underline{q} + \underline{b}_g (\underline{\phi}^T \underline{w} + r) + \tilde{A}_{12} \underline{\eta} \tag{2.51}$$

$$\mu \dot{\underline{\eta}} = A_{22} \underline{\eta} + \mu (\tilde{A}_1 \underline{q} + \underline{A}_2 \underline{\theta}^T \underline{w} + \underline{A}_2 r + A_3 \underline{\eta} + \underline{A}_4 \dot{\underline{\theta}}^T \underline{w} + \underline{A}_4 \underline{\theta}^T \dot{\underline{w}} + \underline{A}_4 \dot{r}) \tag{2.52}$$

It is worth mentioning [15,32,34,55] that when $\underline{\theta} = \underline{\theta}^*$, $\underline{\phi} = \underline{0}$ which makes (2.51) in the parasitic-free case a nonminimal representation of the reference model :

$$\dot{\underline{X}}_n = A_g \underline{X}_n + \underline{b}_g r \tag{2.53}$$

Define the error vector $\underline{e}=\underline{q}-\underline{X}_n$, let $e_1=[1\ 0....0]\underline{e} = \underline{h}_c^T \underline{e}$ and choose

$$\dot{\underline{\theta}} = -\sigma |e_1| \Gamma \underline{\theta} - e_1 \Gamma \underline{w} \tag{2.54}$$

as a rule of adjusting $\underline{\theta}$ with $\sigma>0$ and $\Gamma=\Gamma^T>0$. The resulting adaptive control system with parasitics is described by :

$$\dot{\underline{e}} = A_g \underline{e} + \underline{b}_g \underline{\phi}^T \underline{w} + A_{12} \underline{\eta} \tag{2.55}$$

$$\mu \dot{\underline{\eta}} = A_{22} \underline{\eta} + \mu [\tilde{A}_1 (\underline{e}+\underline{X}_n) + \underline{A}_2 \underline{\theta}^T \underline{w} + \underline{A}_2 r + A_3 \underline{\eta} + \underline{A}_4 \dot{\underline{\theta}}^T \underline{w} + \underline{A}_4 \underline{\theta}^T \dot{\underline{w}} + \underline{A}_4 \dot{r}] \tag{2.56}$$

$$\dot{\underline{\theta}} = - \sigma |e_1| \Gamma \underline{\theta} - e_1 \Gamma \underline{w} \tag{2.57}$$

$$e_1 = \underline{h}_c^T \underline{e} \tag{2.58}$$

<u>Theorem 2.5.1</u>

Let the reference input uniformly bounded; that is

$$|r(t)| \le r_1 \;, \quad |\dot{r}(t)| \le r_2 \quad ; \; r_1 \;, \; r_2 > 0 \quad \forall \quad t \ge 0 \tag{2.59}$$

then there exist positive constant μ^*, t_1, σ, $\alpha < 1/2$, β, γ_1 and c_1-c_4 such that every solution of (2.55)-(2.58) starting at $t = t_0$ from the set

$$\mathcal{R}_{s3}(\mu) = \{(\underline{e},\underline{\theta},\underline{\eta}): \|\underline{e}\| < c_1 \mu^{-\alpha}, \; \|\underline{\theta}\| < c_2 \mu^{-\alpha},$$

$$\|\underline{\eta}\| < c_3 \mu^{-1/2-\alpha} \} \tag{2.60}$$

crosses the target set

$$\mathcal{R}_{t3}(\mu) = \{(\underline{e},\underline{\theta},\underline{\eta}): (\rho_1/8)\|e\|^2 + \beta\|\underline{\theta}-\underline{\theta}^*\|^2 + (\rho_2/8)\|\underline{\eta}\|^2 <$$

$$c_4 \; [(\sigma/4)\|\theta^*\|^4 + \mu^2 \gamma_1 (4(\alpha_0^2 / \rho_1) + (1/\rho_2))]\}$$

$$\tag{2.61}$$

at $t=t_1$ and settles in $\mathcal{R}_{t3}(\mu)$ thereafter. Furthermore,

$$2(\gamma_3/c_1)\, \mu^{2-\alpha} \le \sigma \le \rho_1/2 \quad , \; \beta \le (\sigma \, c_1/2)\, \mu^{-\alpha} \tag{2.62}$$

where ρ_1, ρ_2 and α_0 are some positive constants.

Proof

Consider the $\underline{\Delta}$-parameterized function

$$V(\underline{e},\underline{\theta},\underline{\eta},\underline{\Delta}) = 1/2 \; \underline{e}^T P_1 \underline{e} + 1/2 \; (\underline{\theta}-\underline{\theta}^*)^T \Gamma^{-1} (\underline{\theta}-\underline{\theta}^*) +$$

$$(\mu/2)(\underline{\eta}+\underline{\Delta})^T P_2 (\underline{\eta}+\underline{\Delta}) \tag{2.63}$$

where $P_1 = P_1^T > 0$ satisfies

$$A_g^T P_1 + P_1 A_g = -Q_1 \quad ; \quad Q_1 = Q_1^T > 0 \tag{2.64}$$

$$P_1 \, \underline{b}_g = \underline{h}_c \tag{2.65}$$

and $P_2 = P_2^T > 0$ satisfies

$$A_{22}^T P_2 + P_2 A_{22} = - Q_2 \quad ; \quad Q_2 = Q_2^T > 0 \tag{2.66}$$

Again, one can see that for μ, $c_0 > 0$, $\alpha < 1/2$ and $\underline{\Delta} = - P_2^{-1} A_{22}^{-1^T} \tilde{A}_{21}^T P_1 \underline{e}$, the equality $V = c_0 \mu^{-2\alpha}$ characterizes a closed surface $\mathcal{S}(c_0,\mu,\alpha)$ in the composite space R^{5n+m-3}. The time derivative of (2.63) along the trajectories (2.55)-(2.58), with the aid of (2.64)-(2.66), is given by :

$$\begin{aligned}\dot{V}(\underline{e},\underline{\theta},\underline{\eta}) = {} & -1/2 \, \underline{e}^T Q_1 \underline{e} - \sigma |e_1| \underline{\phi}^T \underline{\theta} - 1/2 \, \underline{\eta}^T Q_2 \underline{\eta} + \mu \, [\underline{\eta} - \\ & - P_2^{-1} A_{22}^{-1^T} \tilde{A}_{12}^T P_1 \underline{e}] P_2 \{ \tilde{A}_1 (\underline{e}+\underline{X}_n) + \underline{A}_2 \underline{\theta}^T \underline{w} + \underline{A}_2 r + A_3 \underline{\eta} \\ & + \underline{A}_4 \dot{\underline{\theta}}^T \underline{w} + \underline{A}_4 \underline{\theta}^T \dot{\underline{w}} + \underline{A}_4 \dot{r} - P_2^{-1} A_{22}^{-1^T} \tilde{A}_{12}^T P_1 [A_g \underline{e} + \\ & + \underline{b}_g \underline{\phi}^T \underline{w} + \tilde{A}_{12} \underline{\eta}]\}\end{aligned} \tag{2.67}$$

Define

$$\rho_1 = \min \lambda(Q_1) \quad , \quad \rho_2 = \min \lambda(Q_2) \tag{2.68}$$

and noting that

$$\|\underline{w}\| \le K_0 \|\underline{e}\| \quad ; \quad K_0 > 0 \tag{2.69}$$

Then (2.67) can be written as

$$\begin{aligned}\dot{V}(\underline{e},\underline{\theta},\underline{\eta}) \le {} & - \rho_1 \|\underline{e}\|^2 - \sigma \, |e_1| \, \|\underline{\phi}\|^2 + \sigma \, |e_1| \, \|\underline{\phi}\| \, \|\underline{\theta}^*\| - \\ & - \rho_2 \|\underline{\eta}\|^2 + \mu (\|\underline{\eta}\| + \alpha_0 \|\underline{e}\|) \{ \alpha_1 \|\underline{e}\|^3 + \alpha_2 \|\underline{e}\|^2 +\end{aligned}$$

$$+ \alpha_3\|\underline{e}\| + \alpha_4\|\underline{\theta}\|^2\,\|\underline{e}\| + \alpha_5\|\underline{\theta}\|\,\|\underline{e}\| + \alpha_6\|\underline{\theta}\|\,\|\underline{e}\|^2 +$$

$$+ \alpha_7\|\underline{\theta}\|^2 + \alpha_8\|\underline{\theta}\| + \alpha_9\|\underline{\theta}\|\,\|\underline{\eta}\| + \alpha_{10}\|\underline{\eta}\| + \gamma_1 \} \qquad (2.70)$$

where γ_1, $\alpha_0 - \alpha_{10} > 0$ determined from the norms of the system matrices. Note that $\gamma_1 = 0$ when $r_1 = r_2 = 0$. For all sequences $(\underline{e}, \underline{\theta}, \underline{\eta})$ originated in $\mathcal{R}_{s3}(\mu)$ with $\sigma > 0$, $|r| < |r_1|$, $|r| < |r_2|$ there exists a μ^* such that for each $\mu \in (0, \mu^*]$, (2.70) becomes

$$\dot{V}(\underline{e}, \underline{\theta}, \underline{\eta}) \le -\|\underline{e}\|^2 \{ \rho_1/4 - \mu\,\alpha_0 [\alpha_1\|\underline{e}\|^2 + \alpha_2\|\underline{e}\| + \alpha_3 + \alpha_4\|\underline{\theta}\|^2$$

$$+ \alpha_5\|\underline{\theta}\| + \alpha_6\|\underline{\theta}\|\,\|\underline{e}\|] - \mu^2\,[\alpha_1\|\underline{e}\|^2 + \alpha_2\|\underline{e}\| + \alpha_3 +$$

$$+ \alpha_4\|\underline{\theta}\|^2 + \alpha_5\|\underline{\theta}\| + \alpha_6\|\underline{\theta}\|\,\|\underline{e}\| + \alpha_0\,\alpha_3\|\underline{\theta}\| + \alpha_0\,\alpha_{10}]^2 \}$$

$$- \rho_2/4\,\{\|\underline{\eta}\| - (2\mu\,/\,\rho_2)\|e\|[\alpha_1\|\underline{e}\|^2 + \alpha_2\|\underline{e}\| + \alpha_3 +$$

$$+ \alpha_4\|\underline{\theta}\|^2 + \alpha_5\|\underline{\theta}\| + \alpha_6\|\underline{\theta}\|\,\|\underline{e}\| + \alpha_0\,\alpha_3\|\underline{\theta}\| + \alpha_0\,\alpha_{10}]\}^2$$

$$- \rho_2/4\,\{\|\underline{\eta}\| - (2\,\mu\,\gamma_1/\rho_2)\}^2 - \|\underline{\eta}\|^2\,\{\rho_2/8 - \mu(\alpha_9 \cdot$$

$$\cdot \|\underline{\theta}\| + \alpha_{10})\} - \rho_1/4\,\{\|\underline{e}\| - (4\,\mu/\rho_1)\,\alpha_0^2\,\|\underline{\theta}\|\,(\alpha_7\|\underline{\theta}\| +$$

$$+ \alpha_8)\}^2 - \rho_1/4\,\{\|\underline{e}\| - (4\,\mu\alpha_0\,\gamma_1/\rho_1)\}^2 - (1/8)(\rho_1 - 2\sigma) \cdot$$

$$\cdot \|\underline{e}\|^2 - (1/2)\|\underline{\phi}\|^2\,[\sigma\|\underline{e}\| - 2\beta] + \mu^2\,\gamma_1\,[(4\alpha_0^2/\rho_1) +$$

$$+ (1/\rho_2)] - \|\underline{\theta}\|^2\,\{(\sigma/2)\|\underline{e}\| - (4\mu^2\,\alpha_0^2/\rho_1)(\alpha_7\|\underline{\theta}\| + \alpha_8)^2\}$$

$$- (\rho_1/8)\|\underline{e}\|^2 - \beta\|\underline{\phi}\|^2 - (\rho_2/8)\|\underline{\eta}\|^2 + (\sigma/4)\|\underline{\theta}^*\|^4$$

$$(2.71)$$

For all sequences $\{\underline{e}(t)\},\{\underline{\phi}(t)\},\{\underline{\eta}(t)\}$ inside $\mathcal{S}(c_0,\mu,\alpha)$, (2.71) is simplified to

$$\dot{V}(\underline{e},\underline{\theta},\underline{\eta}) < -\|\underline{e}\|^2 \{\rho_1/4 - \delta_1 \mu^{1-2\alpha} - \delta_2 \mu^{2(1-2\alpha)}\} - \|\underline{\eta}\|^2 \cdot$$

$$\cdot\{\rho_2/8 - \alpha_{10} \mu - \alpha_9 c_2 \mu^{1-\alpha}\} - (1/8)(\rho_1 - 2\sigma)\|\underline{e}\|^2 -$$

$$- \|\underline{\phi}\|^2\{\sigma c_1 \mu^{-\alpha} - 2\beta\}/2 - (\rho_1/8)\|\underline{e}\|^2 - \beta\|\underline{\phi}\|^2 -$$

$$-(\rho_2/8)\|\underline{\eta}\|^2 + (\sigma/4)\|\underline{\theta}^*\|^4 - \|\underline{\theta}\|^2\{(\sigma c_1/2)\mu^{-\alpha} -$$

$$- \delta_3 \mu^{2(1-\alpha)}\} + \mu^2 \gamma_1 [(4\alpha_0^2/\rho_1) + (1/\rho_2)] \qquad (2.72)$$

for some positive constants δ_1, δ_2 and δ_3. Now, we choose σ and β as in (2.62) and $\alpha<1/2$, there exists a μ^* such that for each $\mu \in (0, \mu^*]$

$$\rho_1/4 \geq \mu^{1-2\alpha}(\delta_1 + \delta_2 \mu^{1-2\alpha}) \text{ and } \rho_2/8 \geq \alpha_9 c_2 \mu^{1-\alpha} + \alpha_{10} \mu \qquad (2.73)$$

Hence, for each $\mu \in (0,\mu^*]$ and all $\underline{e},\underline{\theta},\underline{\eta}$ inside $\mathcal{S}(c_0,\mu,\alpha)$

$$\dot{V}(\underline{e},\underline{\theta},\underline{\eta}) < -(\rho_1/8)\|\underline{e}\|^2 - \beta\|\underline{\phi}\|^2 - (\rho_2/8)\|\underline{\eta}\|^2 + (\sigma/4)\|\underline{\theta}^*\|^4$$

$$+ \mu^2 \gamma_1 [(4\alpha_0^2/\rho_1) + (1/\rho_2)] \qquad (2.74)$$

It is readily evident that there exist constants c_1 to c_3 such that the solutions of (2.55)-(2.58) starting from $\mathcal{R}_{s3}(\mu)$ enter the target set $\mathcal{R}_{t3}(\mu)$ and such that $\mathcal{R}_{t3}(\mu) \subset \mathcal{R}_{s3}(\mu) \subset \mathcal{S}(c_0,\mu,\alpha)$. Note that $\dot{V}(\underline{e},\underline{\theta},\underline{\eta}) < 0$ everywhere in $\mathcal{S}(c_0,\mu,\alpha)$, is monotonically nondecreasing and $V(\underline{e},\underline{\theta},\underline{\eta})$ is strictly decreasing. Therefore, there exist constants $c_4 \geq 1$ and $t_1 \geq t_0$ such that any sequence starting at $t=t_0$ from $\mathcal{R}_{s3}(\mu)$ crosses $\mathcal{R}_{t3}(\mu)$ at $t=t_1$ and settles there for $t>t_1$. Since $r(t)$ is bounded, thus as t increases $\dot{\underline{\phi}}(t) \longrightarrow \underline{0}$,

$\underline{e}(t) \longrightarrow \underline{0}$ and $\underline{\phi}(t) \longrightarrow$ constant and the Theorem is proved.

■■■

It must be emphasized that Theorem 2.5.1 establishes that response of the adaptive control system with parasitics is bounded for bounded input. An important version of Theorem 2.5.1 is given below.

Corollary 2.5.1

For the case of adaptive regulation ($r(t)=0$, $\sigma=0$, $\underline{X}_n(t)=\underline{0}$), there exists a μ^* such that for each $\mu \in (0, \mu^*]$ any solution of (2.55)-(2.58) starting from $\mathcal{R}_{s3}(\mu)$ in (2.60) is bounded as t progress indefinitely.

Proof

The proof follows directly from the proof of the Theorem 2.5.1 by noting that when $r(t)=0$, $\underline{X}_n(t)=\underline{0}$ and $\sigma=0$, the target set $\mathcal{R}_{t3}(\mu)$ reduces to the origin ($\underline{e} = \underline{\eta} = \underline{\theta} = \underline{0}$). Therefore, $\dot{V} \le 0$ everywhere inside $\mathcal{S}(c_0, \mu, \alpha)$ and $V=0$ at the origin of $R^{(5n+m-3)}$. Hence, any solution starts from $\mathcal{R}_{s3}(\mu)$ is bounded and in view of the uniform behaviour of $\dot{V}$ we conclude that $\lim_{t \to \infty} \dot{V}(\underline{e}, \underline{\theta}, \underline{\eta}) = 0$ which implies that

$$\lim_{t \to \infty} \|\underline{e}\| = \lim_{t \to \infty} \|\underline{\eta}\| = 0, \quad \lim_{t \to \infty} \|\dot{\underline{\phi}}\| = 0. \text{ Thus, } \lim_{t \to \infty} \|\underline{\phi}\| =$$

constant and the proof is completed.

■■■

We emphasized that the target set $\mathcal{R}_{t3}(\mu)$ depends on the design parameters (σ, G and $\underline{g}$), reference input characteristics (amplitude and frequency) and frequency range of parasitics.

2.6 Conclusions

This Chapter has dealt with the analysis of reduced-order adaptive control for linear, continuous-time plants in the presence of unmodeled high frequency dynamics. The output error e_1 plays a dual role in the adaptive law and ensures, for sufficiently general bounded input signals, boundedness of both output and parameter errors but within a conservative region of attraction. This region contains all signals which will converge to a target set around the equilibrium point for perfect adaptive tracking. Simulation results indicate that the design parameters can be adjusted to achieve good performance under various operating conditions. The extensions of these results for discrete-time adaptive control systems having external bounded disturbances will be discussed in the next Chapter.

CHAPTER 3

ROBUST CONTROL OF DISCRETE SYSTEMS

3.1 Introduction

Recently, global convergence of adaptively controlled system was resolved for both continuous and discrete-time systems [15-17,21]. However, the most unrealistic among these schemes are the assumptions that there are no disturbances and the plant order is not higher than the order of the model [25,27,53]. It has been shown in [27,53] that the introduction of disturbances, even very small, into the system could deteriorate the efficiency of the control to the point of leading to the instability of the closed-loop system. Furthermore, the study of model reference adaptive schemes in the presence of singular perturbations has been made by Rohrs et al [25,26] which examines situations when the order of the model is equal to the order of the slow part of the unknown plant and the model-plant "mismatch" is due to the fast part "parasitics" of the plant. As shown in [25,35] the disturbances cause a drift of feedback gains to large values and the resulting high gain system is unstable due to the high frequency parasitics. It is then natural to ask how the unknown system will adaptively controlled in the presence of bounded disturbances and/or unmodeled dynamics.

Among the many modifications which have been suggested for adaptive control of such systems, three have gained wide acceptance [28-35,58]. In the first modification [28,30] a dead zone is used in the adaptive law to assure boundedness of all signals in the adaptive loop. A second modification suggested in [27,29] restricts the search region in parameter space by using prior information regarding bounds on the desired control vector $\underline{\theta}^*$. A third modification due to

Ioannou and Kokotovic [32-35], generally referred to as σ-modification, introduces a decaying term , $-\sigma\underline{\theta}$; $\sigma>0$, in the adaptive law for adjusting the parameter vector $\underline{\theta}$. The importance of this latter modification is that no a priori knowledge is required to design the parameter σ. However, as indicated in [31], this approach gives nonzero tracking behaviour in the ideal case (free disturbances and free parasitics). Moreover, it was pointed out in [56,57] that what is called pursting phenomena can occur slowly due to the decaying term , $-\sigma\underline{\theta}$, and resulting in unbounded behaviour or even complete instability of the system. Later on , the authors in [58] replaced the constant σ by a term, $\sigma|e_1|$, where e_1 is the output error. This remodification gave more efficient results.

In the context of discrete-time adaptive control systems , such results can not be readily extended by direct discretization of the pertinent equations [55,59]. It was also pointed out that in simulating continuous systems on a digital computer, there often appear serious convergence problems of the adaptive schemes designed for the continuous case. The above difficulty can be completely eliminated if the continuous system is identified as if it was a discrete one. Consequently,the problem of model reference in discrete-time systems, incorporating bounded disturbances and/or unmodeled dynamics, is of interest in its own right.

In this chapter, the pursued analysis allow simultaneous presence of bounded disturbances, strongly observable parasitics and discrete-plant with throughput. It should be emphasized that, our work contributes to the further development of adaptive control schemes for the case of strictly positive real transfer function of the reference model [55-58]. So, we begin in the next section with the statement of the problem and control objective. In section 3.3 we consider the design of σ-modification approach for a discrete plant with external bounded disturbances , strongly observable parasitics and throughputs. To avoid the instability which may occur due to pursting phenomena, two-modifications are proposed in section 3.4. The main concern in the case of adaptive tracking is that the proposed schemes guarantee boundedness solutions for the system at hand. An estimate of the region of convergence is given in the case of adaptive regulation. Finally, our results are demonstrated by simulating an example of fifth order in section 3.5. It has been shown that the adaptive gain, the mode separation ratio, the adaptation mechanism and the magnitude and periodicity of the reference input sequence are important factors in the design of stable adaptive control schemes.

3.2 SISO Plant With Fast Parasitics and Bounded Disturbances

A general model that describes a linear, discrete-time plants possessing a two-time scale property is given by [55]

$$\underline{X}_s(k+1) = A_1\underline{X}_s(k) + \mu A_2\underline{Z}(k) + \underline{b}_{11}u(k) + \underline{d}_1 \quad , \ |A_1| \neq 0 \qquad (3.1)$$

$$\underline{Z}(k+1)=A_3 \ \underline{X}_s(k) + \mu A_{22}\underline{Z}(k) + \underline{b}_2u(k) + \underline{d}_2 \quad ,| (A_{22})|<1 \qquad (3.2)$$

$$y_s(k) = \underline{C}_1^T \ \underline{X}_s(k) + \mu \ \underline{C}_2^T \ \underline{Z}(k) + d_3 \qquad (3.3)$$

where $\underline{X}_s(k)$, $\underline{Z}(k)$ are the n-dominant and m-parasitic state vectors; respectively, $\mu > 0$ is a scalar parameter, u(k) and y(k) are the scalar input and output; respectively, and $\underline{d}_1$, $\underline{d}_2$ and d_3 are bounded disturbances. Discussion of algorithms to put two-time-scale systems in standard form of type (3.1)-(3.3) are found in [60-64]. It should be emphasized that destabilizing effects can arise in adaptive control when disturbances and/or parasitics are present [25,26,32,34,35]. In order to avoid some instability effects, which are discussed in [35], let us pass the output $y_s(k)$ through a first-order low-pass filter in the form :

$$y_1(k+1) = a\, y_1(k) + y_s(k) \quad ; \quad |a| < 1 \tag{3.4}$$

Now, define $\underline{X}(k) = [y_1(k) \quad \underline{X}_s^T(k)]^T$ and obtain the augmented (n+m+1)th-order plant :

$$\underline{X}(k+1) = A_{11}\underline{X}(k) + \mu\, A_{12}\underline{Z}(k) + \underline{b}_1 u(k) + \underline{D}_1 \tag{3.5}$$

$$\underline{Z}(k+1) = A_{21}\underline{X}(k) + \mu\, A_{22}\underline{Z}(k) + \underline{b}_2 u(k) + \underline{D}_2 \tag{3.6}$$

$$y(k) = y_1(k) = (1\ 0....0)\ \underline{X}(k)$$

$$= \underline{C}_0^T\ \underline{X}(k) \tag{3.7}$$

where $A_{11}, A_{12}, \underline{b}_1, A_{21}$, and the disturbance vectors $\underline{D}_1, \underline{D}_2$ are appropriately defined. We consider the reduced-order adaptive control problem of discrete plant (3.5)-(3.7). The output $y(k)$ is required to track the output $y_m(k)$ of an (n+1)th-order reference model :

$$\underline{X}_m(k+1) = A_m\ \underline{X}_m(k) + \underline{b}_m\ r(k) \tag{3.8}$$

$$y_m(k) = \underline{C}_m^T\ \underline{X}_m(k) + d_m\ r(k) \tag{3.9}$$

whose discrete transfer function is :

$$W_m(z) = d_m + \underline{C}_m^T\ (zI-A_m)^{-1}\ \underline{b}_m = K_m\ N_m(z)/D_m(z) \tag{3.10}$$

where $D_m(z)$ and $N_m(z)$ are monic stable polynomials of degree n+1 and p_m respectively and K_m is a constant. We assume that $W_m(z)$ is strictly positive real [34,59] and $r(k)$ a uniformly bounded signal; that is $|r(k)| \le r_1$ and $|r(k+1)-r(k)| \le r_2$ for all k. The ideal plant

$$\underline{X}_0(k+1) = A_0\ \underline{X}_0(k) + \underline{b}_0\ u(k) \tag{3.11}$$

$$y(k) = \underline{C}_0^T\ \underline{X}_0(k) = (1\ \ 0......0)\ \underline{X}_0(k) \tag{3.12}$$

where $A_o = A_{11} + \mu A_{12}(I - \mu A_{22})^{-1} A_{21}$, $\underline{b}_o = \underline{b}_1 + \mu A_{12}(I - \mu A_{22})^{-1} \underline{b}_2$, obtained by setting $\underline{Z}(k+1)=\underline{Z}(k)$ and $\underline{D}_1=\underline{D}_2=\underline{0}$ in (3.5)-(3.7) is assumed to satisfy the following conditions :

(i) The triple $\{A_o, \underline{b}_o, \underline{C}_o^T\}$ is both completely reachable and completely observable.

(ii) The transfer function

$$W_o(z) = \underline{C}_o^T (zI - A_o)^{-1} b_o = K_p N_o(z)/D_o(z) \tag{3.13}$$

is proper with $D_o(z)$ a monic polynomial of degree n+1, $N_o(z)$ a monic stable polynomial of degree $p_o \le n+1$ and K_p is a constant gain parameter [34,59]. With reference to (3.10) we assume that $p_m \le p_o$; that is the relative degree of the model (3.10) is greater than or equal to that of the plant (3.11).

The controller structure proposed in [15,32,55] is utilized here in conjunction with the representation (3.5)-(3.7). In this structure, the input sequence u(k) and output sequence y(k) are processed to generate two n-dimensional auxiliary vectors $\underline{f}(k)$, $\underline{h}(k)$ in the form :

$$\underline{f}(k+1) = G\,\underline{f}(k) + \underline{g}\,u(k)\ ,\quad \tilde{v}(k) = \underline{C}^T(k)\,\underline{f}(k) \tag{3.14}$$

$$\underline{h}(k+1) = G\underline{h}(k) + \underline{g}y(k)\ ,\quad \tilde{w}(k) = t(k)y(k) + \underline{d}^T(k)\underline{h}(k) \tag{3.15}$$

where G is an nxn stable matrix and the pair $(G,\underline{g})$ is reachable. The input sequence u(k) is given by :

$$u(k) = \underline{\theta}^T(k)\,\underline{w}(k) \tag{3.16}$$

where $\underline{\theta}(k) = [t_o(k), \underline{C}^T(k), t(k), \underline{d}^T(k)]^T$ is a (2n+2)-vector of adjustable parameter and $\underline{w}(k) = [r(k), \underline{f}^T(k), y(k), \underline{h}^T(k)]^T$ is an augmented state vector of order (2n+2). One important feature of the controller structure (3.14)-(3.16) is [34,55] that a constant vector $\underline{\theta}^*$ exists such that for $\underline{\theta}(k) = \underline{\theta}^*$ the transfer function of the ideal plant (3.13) with the controller (3.16) matches that of the model (3.10).

Introducing $\underline{q}^T(k) = [\underline{X}^T(k), \underline{f}^T(k), \underline{h}^T(k)]$, $A_v^T = [A_{12}^T, 0, 0]$,

$\underline{b}_g^T = [\underline{b}_o^T, \underline{g}^T, 0]$, $A_w = [A_{21}, 0, 0]$,

$\underline{D}_c^T = [\underline{D}_1^T + \mu \underline{D}_2^T (I-\mu A_{22})^{-T} A_{12}^T, 0, 0]$ and

$$A_g = \begin{bmatrix} A_o + \underline{b}_o t^* \underline{C}_o^T & \underline{b}_o \underline{C}^{*T} & \underline{b}_o \underline{d}^{*T} \\ \underline{g} t^* \underline{C}_o^T & G + \underline{g}\underline{C}^{*T} & \underline{g}\underline{d}^{*T} \\ \underline{g}\underline{C}_o^T & 0 & G \end{bmatrix} \qquad (3.17)$$

We now combine the controller (3.14)-(3.16) with the discrete plant (3.5)-(3.7) to obtain the augmented system :

$$\underline{q}(k+1) = A_g\, \underline{q}(k) + \underline{b}_g\, [t^* r(k) + \underline{\phi}^T(k)\underline{w}(k)] + \mu\, A_v\, \{\underline{Z}(k) -$$

$$- (I-\mu\, A_{22})^{-1} [A_w \underline{q}(k) + \underline{b}_2 \underline{\theta}^T(k)\underline{w}(k) + \underline{D}_2]\} + \underline{D}_c \qquad (3.18)$$

$$\underline{Z}(k+1) = A_w \underline{q}(k) + \mu\, A_{22}\, \underline{Z}(k) + \underline{b}_2 \underline{\theta}^T(k)\underline{w}(k) + \underline{D}_2 \qquad (3.19)$$

$$y(k) = \underline{h}_c^T\, \underline{q}(k) = [1 \ \ 0.\dots.0]\, \underline{q}(k) \qquad (3.20)$$

where $t^* = K_m/K_p$ and $\underline{\phi} = \underline{\theta}(k) - \underline{\theta}^*$

It can be shown, as in [34,55], that

$$\underline{h}_c^T\ (zI - A_g)^{-1}\ \underline{b}_g = (1/t^*)\ W_m(z)$$

Therefore, using this equation, we can write the following nonminimal representation for the reference model

$$\underline{X}_n(k+1) = A_g \underline{X}_n(k) + \underline{b}_g t^* \, r(k) \tag{3.21a}$$

$$y_m(k) = \underline{h}_c^T \, \underline{X}_n(k) \tag{3.21b}$$

which is used for analysis only and not needed for implementation [34,55].

Define the error vector $\underline{e}(k)=\underline{q}(k)-\underline{X}_n(k)$, let $e_1 = \underline{h}_c^T \underline{e}(k)$ and choose

$$\underline{\theta}(k+1) = \sigma \, H \, \underline{\theta}(k) - e_1 \, H \, \underline{w}(k) \quad ; \tag{3.22a}$$

$$\underline{\phi}(k+1) = \sigma \, H \, \underline{\phi}(k) - e_1 \, H \, \underline{w}(k) + (\sigma H - I) \, \underline{\theta}^* \tag{3.22b}$$

as an adjustment rule for updating the parameter vector $\underline{\theta}(k)$ with a scalar $\sigma > 0$, a matrix $H > 0$ and all eigenvalues of σH exist in the unit circle. As a result, the equations of the error e(k) can be expressed as :

$$\underline{e}(k+1) = A_g \, \underline{q}(k) + \underline{b}_g \, \underline{\phi}^T(k)\underline{w}(k)] + \mu \, A_v \, \{\underline{Z}(k) - (I-\mu A_{22})^{-1} .$$

$$\cdot \, [A_w(\underline{e}(k)+\underline{X}_n(k)) + \underline{b}_2 \underline{\theta}^T(k)\underline{w}(k) + \underline{D}_2]\} + \underline{D}_c \tag{3.23}$$

$$\underline{Z}(k+1) = A_w[\underline{e}(k)+\underline{X}_n(k)] + \mu A_{22} \underline{Z}(k) + \underline{b}_2 \underline{\theta}^T(k)\underline{w}(k) + \underline{D}_2 \tag{3.24}$$

$$e_1(k) = \underline{h}_c^T \, \underline{e}(k) \tag{3.25}$$

We note that an adaptive control system, in the presence of parasitics and bounded disturbances, is now formed by (3.22)-(3.25) for which the stability is the main concern. The stability properties are analyzed in the next sections.

3.3 Adaptive System With Parasitics and Bounded Disturbances

The fast state $\underline{Z}(k)$ in (3.24) is formed of a "fast transient" and a "quasi-steady state" $\underline{\tilde{Z}}$ defined as the solution of (3.24) with $\underline{Z}(k+1)=\underline{Z}(k)$ [55]. This motivates the definition of the fast parasitic state $\underline{\eta}(k)$ as the difference between $\underline{Z}(k)$ and $\underline{\tilde{Z}}$, i.e.

$$\underline{\eta}(k+1) = \underline{Z}(k) - (I-\mu A_{22})^{-1} \{A_w[\underline{e}(k)+\underline{X}_n(k)] + \underline{b}_2\underline{\theta}^T(k)\underline{w}(k) + \underline{D}_2\} \tag{3.26}$$

Using (3.26), we can write (3.22)-(3.25) as

$$\underline{e}(k+1) = A_g\underline{e}(k) + \underline{b}_g\underline{\phi}^T(k)\underline{w}(k) + \mu\, A_v\, \underline{\eta}(k) + \underline{D}_c \tag{3.27}$$

$$\underline{\eta}(k+1) = \mu\, A_{22}\, \underline{\eta}(k) - (I-\mu A_{22})^{-1}\, [\, \underline{\xi}(k+1) - \underline{\xi}(k)] \tag{3.28}$$

$$\underline{\xi}(k) = A_w[\underline{e}(k)+\underline{X}_n(k)] + \underline{b}_2\underline{\theta}^T(k)\underline{w}(k) + \underline{D}_2 \tag{3.29}$$

$$e_1(k) = \underline{h}_c^T\, \underline{e}(k) \tag{3.30}$$

$$\underline{\phi}(k+1) = \sigma\, H\, \underline{\phi}(k) - e_1\, H\, \underline{w}(k) + (\sigma H - I)\, \underline{\theta}^* \tag{3.31}$$

To investigate the stability properties of the adaptive system (3.27)-(3.31), we choose

$$V(\underline{e},\underline{\phi},\underline{\eta}) = \underline{e}^T(k)P_1\underline{e}(k) + \underline{\phi}^T(k)P_2\underline{\phi}(k) + [\,\underline{\eta}(k)+(I-\mu A_{22})^{-1} \cdot \underline{\xi}(k)]^T P_3 [\,\underline{\eta}(k)+(I-\mu A_{22})^{-1}\, \underline{\xi}(k)] \tag{3.32}$$

as a candidate Lyapunov function, where P_1, P_2 and P_3 are symmetric, positive-definite matrices satisfying

$$A_g^T\, P_1\, A_g - P_1 = -Q_1 \quad ; \quad Q_1 = Q_1^T > 0 \tag{3.33}$$

$$\sigma^2\, H^T\, P_2\, H - P_2 = -Q_2 \quad ; \quad Q_2 = Q_2^T > 0 \tag{3.34}$$

$$\mu^2\, A_{22}^T\, P_3\, A_{22} - P_3 = -Q_3 \quad ; \quad Q_3 = Q_3^T > 0 \tag{3.35}$$

It is important to observe that for each $\mu > 0$, $c_0 > 0$ and $\alpha < 1/2$ the relation

$$V(\underline{e},\underline{\phi},\underline{\eta}) = c_0\ \mu^{-2\alpha} \tag{3.36}$$

characterizes a closed surface $\mathcal{S}(c_0,\mu,\alpha)$ in the composite space $R^{(5n+m+3)}$. In the light of (3.36) and μ-parameterization of the adaptive system, we consider for a given μ that the solution of (3.27)-(3.31) starting at $k=k_0$ from the region

$$\mathcal{R}_s(\mu) = \{(\underline{e},\underline{\phi},\underline{\eta}) : \|\underline{e}\| \le c_1\ \mu^{-\alpha},\ \|\underline{\phi}\| \le c_2\ \mu^{-\alpha},$$

$$\|\underline{\eta}\| \le c_3\ \mu^{-\alpha}\} \tag{3.37}$$

are enclosed by $\mathcal{S}(c_0,\mu,\alpha)$, where c_1, c_2 and c_3 are positive constants. It should be emphasized that the equality (3.36) defines a closure of a domain of attraction of the stable solutions. For the parasitic-free case ($\underline{\eta}=\underline{0}$) and without disturbances, the solution $\{\underline{e}(k),\underline{\phi}(k)\}$ of (3.27)-(3.31) is bounded for any bounded initial conditions $\{\underline{e}(0),\underline{\phi}(0)\}$ and bounded input sequences [55]. Let $\lambda_m(.)$, $\lambda_M(.)$ denote the minimum, maximum eigenvalues of matrix (.). The following theorem summarizes the main stability result for the adaptive system (3.27)-(3.31) with parasitics and/or bounded disturbances.

<u>Theorem 3.3.1</u>

Let the input reference uniformly bounded; that is

$$|r(k)| < r_1,\ |r(k+1)-r(k)| < r_2\ ;\ r_1, r_2 > 0\ \forall\ k \ge k_0 \tag{3.38}$$

Then, there exist positive scalars μ^*, σ, $\alpha<1/2$, c_1 to c_3, k_t and c_t such that for each $\mu \in (0, \mu^*]$ every solution of the augmented system (3.27)-(3.31) evolving at $k = k_0$ from the initial region $\mathcal{R}_s(\mu)$ crosses the target set

$$\mathcal{R}_{t1}(\mu) = \{(\underline{e},\underline{\phi},\underline{\eta}) : (\rho_1/2)\|\underline{e}\|^2 + (\rho_2/2)\|\underline{\phi}\|^2 (\rho_3/2)\|\underline{\eta}\|^2$$

$$\leq c_t \ [D_o^2 \ (\lambda_1 + \beta_{11} + \mu \ \beta_{12}) + 3 \ \lambda_2 \ \|\sigma H - I\|^2 \ \|\underline{\theta}^*\|^2$$

$$+ \ \Omega(\mu,\sigma,r)]\} \tag{3.39}$$

at $k=k_t$ and settles in $\mathcal{R}_{t1}(\mu)$ thereafter, where

$D_o = \sup_k \|\underline{D}_c(k)\|$, ρ_1, ρ_2, ρ_3, λ_1, λ_2, β_{11}, β_{12} are positive constants and $\Omega(\mu,\sigma,r)$ is a functional scalar. Furthermore, $\sigma \in F_1(\sigma)$ such that

$$F_1(\sigma) = \{\sigma : 0 \leq \sigma \leq (\rho_2/8 \ \lambda_2)^{1/2} \ \} \tag{3.40}$$

<u>Proof</u>

The first forward difference of $V(\underline{e},\underline{\phi},\underline{\eta})$ along the solution of (3.27)-(3.31) is given by :

$$\Delta V(\underline{e},\underline{\phi},\underline{\eta}) = \{A_g \underline{e}(k) + \underline{b}_g \underline{\phi}^T(k)\underline{w}(k) + \mu A_v \underline{\eta}(k) + \underline{D}_c(k)\}^T \ P_1 \bullet$$

$$\bullet \{A_g \underline{e}(k) + \underline{b}_g \underline{\phi}^T(k)\underline{w}(k) + \mu A_v \underline{\eta}(k) + \underline{D}_c(k)\} +$$

$$+ \ \{\sigma H\underline{\phi}(k) - e_1 H\underline{w}(k) + (\sigma H - I)\underline{\theta}^*\}^T \ P_2 \ \{\sigma H\underline{\phi}(k) -$$

$$- \ e_1 H\underline{w}(k) + (\sigma H - I)\underline{\theta}^*\} + \{ \underline{\eta}(k+1) + (I - \mu A_{22})^{-1} \bullet$$

$$\bullet \underline{\xi}(k+1)\}^T \ P_3 \ \{ \underline{\eta}(k+1) + (I - \mu A_{22})^{-1} \ \underline{\xi}(k+1)\} -$$

$$- \ \underline{e}^T(k) \ P_1 \ \underline{e}(k) - \underline{\phi}^T(k) \ P_2 \ \underline{\phi}(k) - [\ \underline{\eta}(k) +$$

$$+ \ (I - \mu A_{22})^{-1} \ \underline{\xi}(k)]^T \ P_3 \ [\ \underline{\eta}(k) + (I - \mu A_{22})^{-1} \ \underline{\xi}(k)] \tag{3.41}$$

Since the reference input r(k) is uniformly bounded, then there exist positive constants α_0, $\epsilon_0 - \epsilon_3$ such that

$$\|\underline{w}(k)\| \le \alpha_0 + \|\underline{e}\| \tag{3.42a}$$

$$\|\underline{\xi}(k)\| \le \epsilon_0 + \epsilon_1 \|\underline{e}\| + \epsilon_2 \|\underline{\phi}\| + \mu\, \epsilon_3 \|\underline{\eta}\| \tag{3.42b}$$

where the constants α_0, ϵ_0 depend on the bounds for

$|r(k)|, |r(k+1)-r(k)|$ such that $\alpha_0 = \epsilon_0 = 0$ when $r_1 = r_2 = 0$.

Define

$$\rho_1 = \lambda_m(Q_1)\ , \quad \rho_2 = \lambda_m(Q_2)\ , \quad \rho_3 = \lambda_m(Q_3)$$

$$\lambda_1 = \lambda_M(P_1)\ , \quad \lambda_2 = \lambda_M(P_2)\ , \quad \lambda_3 = \|P_3\| \tag{3.43}$$

Algebraic manipulation of (3.41) using (3.42) and (3.43) yields :

$$\begin{aligned}
\Delta V(\underline{e},\underline{\phi},\underline{\eta}) \le\ & -(3\rho_1/4)\|\underline{e}\|^2 + \alpha_0\alpha_1\|\underline{e}\|\,\|\underline{\phi}\| + \alpha_1\|\underline{e}\|^2\|\underline{\phi}\| - \\
& - \rho_1/4\,\{\|\underline{e}\| - (2\mu/\rho_1)\|\underline{\eta}\|(\alpha_2 + \alpha_5\|\underline{\phi}\|)\}^2 + \alpha_3 D_0\|\underline{e}\| + \\
& + \alpha_0^2\,\alpha_4\|\underline{\phi}\|^2 + 2\alpha_0\,\alpha_4\|\underline{e}\|\,\|\underline{\phi}\|^2 + \alpha_4\|\underline{e}\|^2\|\underline{\phi}\|^2 + \\
& + \alpha_0\,\alpha_5\|\underline{\phi}\|\,\|\underline{\eta}\| + \alpha_0\,\alpha_6 D_0\|\underline{\phi}\| + \alpha_6 D_0\|\underline{e}\|\,\|\underline{\phi}\| + \\
& + \mu^2\|\underline{\eta}\|^2[\alpha_7 + (1/\rho_1)(\alpha_2 + \alpha_5\|\underline{\phi}\|)^2] + \mu\alpha_8 D_0\|\underline{\eta}\| + \\
& + \lambda_1 D_0^2 - \rho_2\|\underline{\phi}\|^2 + 2\sigma^2\lambda_2\|\underline{\phi}\|^2 + \alpha_0^2\,\alpha_9\|\underline{e}\|^2 + \\
& + 2\alpha_0\,\alpha_9\|\underline{e}\|^3 + \alpha_9\|\underline{e}\|^4 + 3\lambda_2\|\sigma H - I\|^2\|\underline{\theta}^*\|^2 - \\
& - \lambda_2\|\sigma H\underline{\phi} + e_1 H\underline{w}\|^2 - \lambda_2\|\sigma H\underline{\phi} - (\sigma H - I)\underline{\theta}^*\|^2 -
\end{aligned}$$

$$- \lambda_2 \| e_1 H \underline{w} + (\sigma H - I)\underline{\theta}^* \|^2 - \rho_3 \|\underline{\eta}\|^2 + 2\lambda_3 \epsilon_0 \|\underline{\eta}\| +$$

$$+ \alpha_{10} \|\underline{e}\| \|\underline{\eta}\| + \alpha_{11} \|\underline{\phi}\| \|\underline{\eta}\| + \mu \alpha_{12} \|\underline{\eta}\|^2 \qquad (3.44)$$

for some positive scalars ($\alpha_0 - \alpha_{12}$, ϵ_0), determined from the norm of the system matrices, and α_0, ϵ_0 depend on r_1, r_2.

Now for all sequences $\{\underline{e}(k)\}, \{\underline{\phi}(k)\}, \{\underline{\eta}(k)\}$ originated in $\mathcal{R}_s(\mu)$ we can express (3.44), after grouping terms, as :

$$\Delta V(\underline{e}, \underline{\phi}, \underline{\eta}) \le -\|\underline{e}\|^2 \{ \rho_1/4 - (\alpha_3 + \alpha_{10})/2 - \alpha_1 c_2 \mu^{-\alpha}$$

$$- \alpha_9 c_1^2 \mu^{-2\alpha} \} - \|\underline{\phi}\|^2 \{ \rho_2/4 - 2\sigma^2 \lambda_2 \} - \|\underline{\phi}\|^2 \{ \rho_2/4 -$$

$$- \alpha_{11}/2 - (\alpha_4 + \alpha_6/2) c_1^2 \mu^{-2\alpha} \} - \|\underline{\eta}\|^2 \{ \rho_3/2 -$$

$$- (\alpha_{10} + \alpha_{11})/2 - \mu(\alpha_8/2 + \alpha_{12}) - \mu^2 [\alpha_7 + (1/\rho_1)(\alpha_2$$

$$+ \alpha_5 c_2 \mu^{-\alpha})^2]\} + D_0^2 \{ \lambda_1 + (\alpha_3 + \alpha_6)/2 + \mu\alpha_8/2 \} -$$

$$-(\rho_1/2) \|\underline{e}\|^2 - (\rho_2/2) \|\underline{\phi}\|^2 - (\rho_3/2) \|\underline{\eta}\|^2$$

$$+ 3\lambda_2 \|\sigma H - I\|^2 \|\underline{\theta}^*\|^2 + \Omega(\mu, \sigma, r) \qquad (3.45)$$

where the functional scalar $\Omega(\mu,\sigma,r)$ is given by :

$$\Omega(\mu,\sigma,r) = [\alpha_0 \alpha_6 D_0 c_2 + 2\lambda_3 \epsilon_0 c_3] \mu^{-\alpha} + [\alpha_0 \alpha_1 c_1 c_2 +$$

$$+ \alpha_0^2 \alpha_4 c_2^2 + \alpha_0^2 \alpha_9 c_1^2] \mu^{-2\alpha} + 2\alpha_0 c_1 [\alpha_4 c_2^2 +$$

$$+ \alpha_9 c_1^2] \mu^{-3\alpha} + \alpha_0 \alpha_5 c_2 c_3 \mu^{1-2\alpha} \qquad (3.46)$$

For $\underline{e}, \underline{\phi}, \underline{\eta}$ inside $\mathcal{S}(c_0, \mu, \alpha)$, (3.45) can be further simplified to :

$$\Delta V(\underline{e},\underline{\phi},\underline{\eta}) \le -\|\underline{e}\|^2 \{\rho_1/4 - \beta_1 - \beta_2 \mu^{-\alpha} - \beta_3 \mu^{-2\alpha}\} - \|\underline{\phi}\|^2 \cdot$$

$$\cdot \{\rho_2/4 - 2\sigma^2 \lambda_2\} - \|\underline{\phi}\|^2 \{\rho_2/4 - \beta_4 - \beta_5 \mu^{-2\alpha}\} - \|\underline{\eta}\|^2 \cdot$$

$$\cdot \{\rho_3/2 - \beta_6 - \beta_7 \mu - \beta_8 \mu^2 - \beta_9 \mu^{2-\alpha} - \beta_{10} \mu^{2(1-\alpha)}\} +$$

$$+ D_0^2 [\lambda_1 + \beta_{11} + \mu\beta_{12}] - \rho_1/2 \|\underline{e}\|^2 - \rho_2/2 \|\underline{\phi}\|^2 -$$

$$- \rho_3/2 \|\underline{\eta}\|^2 + 3\lambda_2 \|\sigma H - I\|^2 \|\underline{\theta}^*\|^2 + \Omega(\mu,\sigma,r) \qquad (3.47)$$

where

$$\Omega(\mu,\sigma,r) = \delta_1 \mu^{-\alpha} + \delta_2 \mu^{-2\alpha} + \delta_3 \mu^{-3\alpha} + \delta_4 \mu^{1-2\alpha} \qquad (3.48)$$

for some positive constants β_1-β_{12} and δ_1-δ_4. Note that δ_1-δ_4 depend on the bounds for $|r(k)|$ and $|\Delta r(k)|$ such that $\delta_1 = \delta_2 = \delta_3 = \delta_4 = 0$ when $r_1 = r_2 = 0$

On choosing $0 < \sigma < (\rho_2/8\lambda_2)^{1/2}$, $\alpha < 1/2$, we can find a μ^* such that for each $\mu \in (0, \mu^*]$

$$\rho_1/4 \ge \beta_1 + \beta_2 \mu^{-\alpha} + \beta_3 \mu^{-2\alpha} \qquad (3.49a)$$

$$\rho_2/4 \ge \beta_4 + \beta_5 \mu^{-2\alpha} \qquad (3.49b)$$

$$\rho_3/4 \ge \beta_6 + \beta_7 \mu + \beta_8 \mu^2 + \beta_9 \mu^{2-\alpha} + \beta_{10} \mu^{2(1-\alpha)} \qquad (3.49c)$$

and $\Omega(\mu,\sigma,r) > 0$. Hence,

$$\Delta V < - \rho_1/2 \|\underline{e}\|^2 - \rho_2/2 \|\underline{\phi}\|^2 - \rho_3/2 \|\underline{\eta}\|^2 + D_0^2 [\lambda_1 + \beta_{11}$$

$$+ \mu\beta_{12}] + 3\lambda_2 \|\sigma H - I\|^2 \|\underline{\theta}^*\|^2 + \Omega(\mu,\sigma,r) \qquad (3.50)$$

for all $\mu \in (0,\mu^*]$ and $\underline{e}$, $\underline{\phi}$, $\underline{\eta}$ inside $\mathcal{S}(c_0,\mu,\alpha)$. In view of the boundedness of the input sequence, the functional $\Omega(\mu,\sigma,r)$ is uniformly bounded and therefore $\mathcal{R}_{t1}(\mu)$ is a uniformly bounded region. It is readily evident that there exist constants c_1 to c_3 such that the solutions of (3.27)-(3.31) starting from $\mathcal{R}_s(\mu)$ enter the target set $\mathcal{R}_{t1}(\mu)$ and such that $\mathcal{R}_{t1}(\mu) \subset \mathcal{R}_s(\mu) \subset \mathcal{S}(c_0,\mu,\alpha)$. Observe that $\Delta V(\underline{e},\underline{\phi},\underline{\eta})$ < 0 everywhere in $\mathcal{S}(c_0,\mu,\alpha)$, except possibly in $\mathcal{R}_{t1}(\mu)$, and $V(\underline{e},\underline{\phi},\underline{\eta})$ is monotonically non-increasing in $\mathcal{R}_{t1}(\mu)/\mathcal{R}_s(\mu)$. This means that there exist constants $k_t \geq k_0$, $c_t \geq 1$ such that any solution starting at $k=k_0$ from $\mathcal{R}_s(\mu)$ will cross the target set $\mathcal{R}_{t1}(\mu)$ at $k=k_t$ and settles in $\mathcal{R}_{t1}(\mu)$ for all $k>k_t$. As k increases and in view of boundedness of the reference input and external disturbance, we see that the solutions of (3.27)-(3.31) converge to the residual set $\mathcal{R}_{t1}(\mu)$ which completes the proof. ▪▪▪

It must be emphasized that Theorem 3.3.1 establishes that response of the adaptive control system with bounded disturbances and/or parasitics is bounded for bounded input sequences. An important version of Theorem 3.3.1 is given below.

<u>Corollary 3.3.1</u>

Consider the regulation case $r(k) = 0$, $\sigma H = I$ and $\underline{X}_n(k) = \underline{0}$. The disturbances are absent, that is $\underline{D}_c = \underline{0}$ and $\underline{D}_2 = \underline{0}$ in (3.27) and (3.29). Then, there exist a μ^* such that for each $\mu \in (0,\mu^*]$ any solution $(\underline{e},\underline{\phi},\underline{\eta})$ of the augmented system (3.27)-(3.31) which evolves from $\mathcal{R}_s(\mu)$ in (3.37) is bounded, that is

$$\lim_{k\to\infty} \|\underline{q}(k)\| = \lim_{k\to\infty} \|\underline{e}(k)\| = 0 , \quad \lim_{k\to\infty} \|\underline{\eta}(k)\| = 0 \text{ and}$$

$$\lim_{k\to\infty} \|\underline{\phi}(k+1)\| = 0 \quad \text{and} \quad \lim_{k\to\infty} \|\underline{\theta}(k)-\underline{\theta}^*\|_{P_2} = \text{constant.}$$

Proof

It is straightforward to see in the regulation case that the functional scalar $\Omega(\mu,\sigma,r) = 0$, $D_o = 0$ and $\underline{e}(k) = \underline{q}(k)$. Moreover, for each $\mu \in (0,\mu^*]$, (3.49) is automatically satisfied. It directly follows that $\mathcal{R}_{t1}(\mu)$ in (3.39) reduces to the origin ($\underline{e}=\underline{0}, \underline{\phi}=\underline{0}, \underline{\eta}=\underline{0}$). Therefore, $\Delta V(\underline{e},\underline{\phi},\underline{\eta}) \le 0$ everywhere inside $\mathcal{S}(c_0,\mu,\alpha)$ and $\Delta V(\underline{e},\underline{\phi},\underline{\eta})=0$ at the origin of the composite space R^{5n+m+3}. As a result, any solution starting from $\mathcal{R}_s(\mu)$ is bounded, and by virtue of the uniform boundedness of $\Delta V(\underline{e},\underline{\phi},\underline{\eta})$, it is readily evident that $\lim_{k\to\infty} \Delta V(\underline{e},\underline{\phi},\underline{\eta}) = 0$. Consequently, we have $\lim_{k\to\infty} \|\underline{q}(k)\| = 0$,

$\lim_{k\to\infty} \|\underline{\phi}(k+1)\| = 0$, and $\lim_{k\to\infty} \|\underline{\theta}(k)-\underline{\theta}^*\|_{P_2} = V_\infty$ which is

finite constant. ■■■

3.4 Modified Parameter Adjustment

In the above section, it has been shown that the reduced-order adaptive control system with the adaptation law (3.22) has bounded behaviour and all signals converge to a residual set whose size depends on σ, μ and the external bounded disturbance. However, bursting phenomena can occur slowly due to the decaying term $\sigma H\underline{\theta}(k)$, resulting in sudden intermittent output error "bursts" followed by a long period of the apparent behaviour of the system or even complete instability of the system especially for large values of σ [56,57].

To overcome these undesirable effects new adaptive laws are proposed in which the constant σ in (3.22) is replaced by a nonlinear function $f(e_1)$ satisfying $|f(e_1)| \le 1$, where e_1 is defined above. The proposed modifications will be shown to improve the performance of the system in all aspects while retaining the advantage of (3.22) of assuring robustness in the presence of unmodeled parasitics and/or external bounded disturbances, without requiring additional information regarding the plant or disturbances.

A. Sgn-modification Adaptation Law

The controller parameter $\underline{\theta}(k)$ is updated along the adjustment rule :

$$\underline{\theta}(k+1) = \sigma \,\mathrm{sgn}(e_1)\, H\, \underline{\theta}(k) - e_1 H\, \underline{w}(k) \quad ; \tag{3.51a}$$

$$\underline{\phi}(k+1) = \sigma \mathrm{sgn}(e_1) H \underline{\phi}(k) - e_1 H \underline{w}(k) + [\sigma \mathrm{sgn}(e_1) H - I]\underline{\theta}^* \tag{3.51b}$$

where $e_1(k)$ is given by (3.30) and

$$\mathrm{sgn}(e_1) = \begin{cases} 1 & e_1 > 0 \\ -1 & e_1 \leq 0 \end{cases} \tag{3.52}$$

In this case, the stability properties of the adaptive system (3.27)-(3.30) with sgn-modification adaptation law (3.51), are characterized by the following Theorem :

Theorem 3.4.1

Let the reference input uniformly bounded. There exist positive scalars μ^*, $\alpha<1/2$, c_1 to c_3, k_t and c_t such that if $\sigma \in F_2(\sigma)$ where

$$F_2(\sigma) = \{\sigma : -1/\|H\|^2 \leq \sigma \leq (\rho_2/8\lambda_2)^{1/2}\} \tag{3.53}$$

then for each $\mu \in (0,\mu^*]$ the solution $\underline{e}(k), \underline{\phi}(k), \underline{\eta}(k)$ of (3.27)-(3.30),(3.51) which starts from $\mathcal{R}_s(\mu)$ in (3.37) is

(i) uniformly bounded

(ii) converges to the residual set

$$\mathcal{R}_{t2}(\mu) = \{(\underline{e},\underline{\phi},\underline{\eta}) : \varsigma_1/2 \|\underline{e}\|^2 + \varsigma_2/2 \|\underline{\phi}\|^2 + \varsigma_3/2 \|\underline{\eta}\|^2 \le$$

$$\le ct \ [D_0^2 \ (\lambda_1 + \beta_{11} + \mu\beta_{12}) + 3\lambda_2 \|\sigma H + I\|^2 \|\underline{\theta}^*\|^2 +$$

$$+ \Omega(\mu,\sigma,r)]\} \qquad (3.54)$$

<u>Proof</u>

It can be easily obtained by following analysis similar to (3.41)-(3.50). ▪▪▪

<u>B. Sat-modification Adaptation Law</u>

The adaptive law (3.22) can also be modified to have the form :

$$\underline{\theta}(k+1) = \sigma \ sat(e_1) \ H \ \underline{\theta}(k) - e_1 \ H \ \underline{w}(k) \quad ; \qquad (3.55a)$$

$$\underline{\phi}(k+1) = \sigma \ sat(e_1) H \underline{\phi}(k) - e_1 \ H\underline{w}(k) + [\sigma sat(e_1) H - I]\underline{\theta}^* \qquad (3.55b)$$

where $e_1(k)$ is given by (3.30) and

$$sat(e_1) = \begin{cases} 1 & e_1 > 1 \\ e_1 & |e_1| \le 1 \\ -1 & e_1 < -1 \end{cases} \qquad (3.56)$$

In this case, we have the following result.

<u>Theorem 3.4.2</u>

Let r(k) and r(k) satisfy :

$$|r(k)| < r_1 \ , \ |r(k+1) - r(k)| < r_2 \quad ; \ r_1, r_2 > 0 \ \forall \ k \ge k_0 \qquad (3.57)$$

and let $\sigma \in F_3(\sigma)$ where

$$F_3(\sigma) = \{\sigma : -1/(\gamma^2 \|H\|^2) \leq \sigma \leq (\rho_2/8\lambda_2)^{1/2}\} \quad (3.58)$$

for some positive constant $\gamma \leq 1$. Then, there exist positive scalars μ^*, $\alpha<1/2$, c_1 to c_3, k_t and c_t such that every solution of the augmented system (3.27)-(3.30) with adaptation law (3.55) starting at $k=k_0$ from the $\mathcal{R}_s(\mu)$ enter the target set

$$\mathcal{R}_{t3}(\mu) = \{(\underline{e},\underline{\phi},\underline{\eta}) : (\rho_1/2)\|\underline{e}\|^2 + (\rho_2/2)\|\underline{\phi}\|^2 + (\rho_3/2)\|\underline{\eta}\|^2$$
$$\leq c_t\ [D_0^2\ (\lambda_1 + \beta_{11} + \mu\ \beta_{12}) + 3\lambda_2\|\sigma\gamma H+I\|^2\ \|\underline{\theta}^*\|^2$$
$$+ \Omega(\mu,\sigma,r)]\} \quad (3.59)$$

at $k = k_t$ and remains in $\mathcal{R}_{t3}(\mu)$ for all $k \geq k_t$.

<u>Proof</u>

The Theorem can be easily proved by applying the Lyapunov analysis pursued in (3.41)-(3.50). ▪▪▪

<u>Remark 3.4.1</u>

The two modification schemes (3.51) and (3.55) allow the designer to select the design parameter σ from definite sets $F_2(\sigma)$ in the first modified law and $F_3(\sigma)$ in the second one. Moreover, σ can be chosen to be positive or negative value as shown in (3.53) and (3.58).

<u>Remark 3.4.2</u>

In the regulation case ($r(k)=0$, $\underline{X}_n(k)=\underline{0}$ and $\sigma H=-I$ in the sgn-modification or $\sigma\ H=-I$ in the sat-modification) and in the disturbance free case ($\underline{d}_1=\underline{d}_2=d_3=0$ in (3.1)-(3.3)), it follows easily from (3.54) and (3.59) that $\mathcal{R}_{t2}(\mu)$ and

$\mathcal{R}_{t3}(\mu)$ reduce to the origin and hence the domain of stability in the modified adaptive laws (3.51), (3.55) will represent the whole set $\mathcal{S}(c_0,\mu,\alpha)$. As a result, there exist a μ^* such that for each $\mu \in (0, \mu^*)$ any solution of (3.27)-(3.30), with (3.51) or (3.55), starting from $\mathcal{R}_s(\mu)$ in (3.37) is bounded as k progresses indefinitely.

Remark 3.4.3

It should be noted that the assumption $\mu \in (0,\mu^*]$ has been used throughout our work. It has been shown in [65] that there exist a $\mu_0 > 0$, $\mu^* > \mu_0$ such that constant output feedback laws of the form $u(k) = \underline{\theta}_0^T \underline{y}(k)$ stabilizes (3.18)-(3.20) for all $\mu \in (0, \mu_0]$. The scalar μ_0 is expressed in terms of norms of the system matrices.

Remark 3.4.4

It should be emphasized that the functional scalar $\Omega(\mu,\sigma,r)$ depends on $|r|=r_1$ and $|\ r|=r_2$. This shows that large magnitude or rapid variation of a reference input sequence no longer guarantee that $\Delta V < 0$ everywhere in $\mathcal{R}_t(\mu)/\mathcal{R}_s(\mu)$. The main reason is that such reference sequences adds to the control input sequences which lie in the parasitic range. this in turn excites the parasitic modes and consequently leads to system instability. Such phenomena has been observed before in adaptive control schemes [32-35,67,68].

3.5 ILLUSTRATIVE EXAMPLE

We now demonstrate the robustness properties of the model reference adaptive schemes by digital simulation for a plant of type (3.1)-(3.3) with n=2, m=3 and

$$A_1 = \begin{bmatrix} -.3212 & .1927 \\ .1134 & -.4195 \end{bmatrix}, \quad A_2 = \begin{bmatrix} .1436 & .0567 & .1436 \\ -.0019 & .1739 & .4195 \end{bmatrix}$$

$$A_3 = \begin{bmatrix} -.006 & .468 \\ -.715 & -.022 \\ -.148 & -.003 \end{bmatrix} , \quad A_{22} = \begin{bmatrix} .247 & .014 & .048 \\ -.021 & .24 & -.024 \\ -.004 & .09 & .026 \end{bmatrix}$$

$$\underline{b}_{11}^T = [.037 \quad .4611] , \qquad \underline{b}_2^T = [.036 \quad .562 \quad .115]$$

$$\underline{d}_1^T = [0.1 \quad 0.3] , \quad \underline{d}_2^T = [0.2 \quad 0.4 \quad 0.1] , \quad d_3 = 0.5$$

$$\underline{C}_1^T = [1.0 \quad 1.0] , \quad \underline{C}_2^T = [1.0 \quad 1.0 \quad 1.0]$$

$$\underline{X}^T(0) = [0.5 \quad 0.8] , \quad \underline{Z}^T(0) = [0.4 \quad 0.5 \quad 0.8]$$

The low-pass filter (3.4) is given by :

$$y_1(k+1) = 0.2\, y_1(k) + y_s(k) , \quad y_1(0) = 1.0$$

and the reference model (3.8),(3.9) is specified by :

$$A_m = \begin{bmatrix} .1 & .2 & .5 \\ -.2 & .34 & .4 \\ -.3 & .1 & -.2 \end{bmatrix} , \quad \underline{b}_m = \begin{bmatrix} 0.1 \\ 0.2 \\ 0.1 \end{bmatrix} , \quad d_m = [\ 0.01\ ]$$

$$\underline{C}_m^T = [1.0 \quad 0.0 \quad 0.0] , \quad \underline{X}_m^T(0) = [0.1 \quad -0.3 \quad 0.4]$$

Other parameters are

$$G = \begin{bmatrix} .5 & 0 \\ 0 & .3 \end{bmatrix} , \quad \underline{g} = \begin{bmatrix} .1 \\ .2 \end{bmatrix} , \quad \underline{f}(0) = \begin{bmatrix} 1 \\ .5 \end{bmatrix} , \quad \underline{h}(0) = \begin{bmatrix} -.5 \\ .1 \end{bmatrix}$$

and $H = I_6$. The three adaptation laws (3.22), (3.51) and (3.55) are carried out and the simulation results are summarized in Figs. 3.1-3.12 due to different values of σ, μ and magnitude and priodicity of the reference input. In Figs. 3.1-3.3, the error signal e_1 is plotted against time for

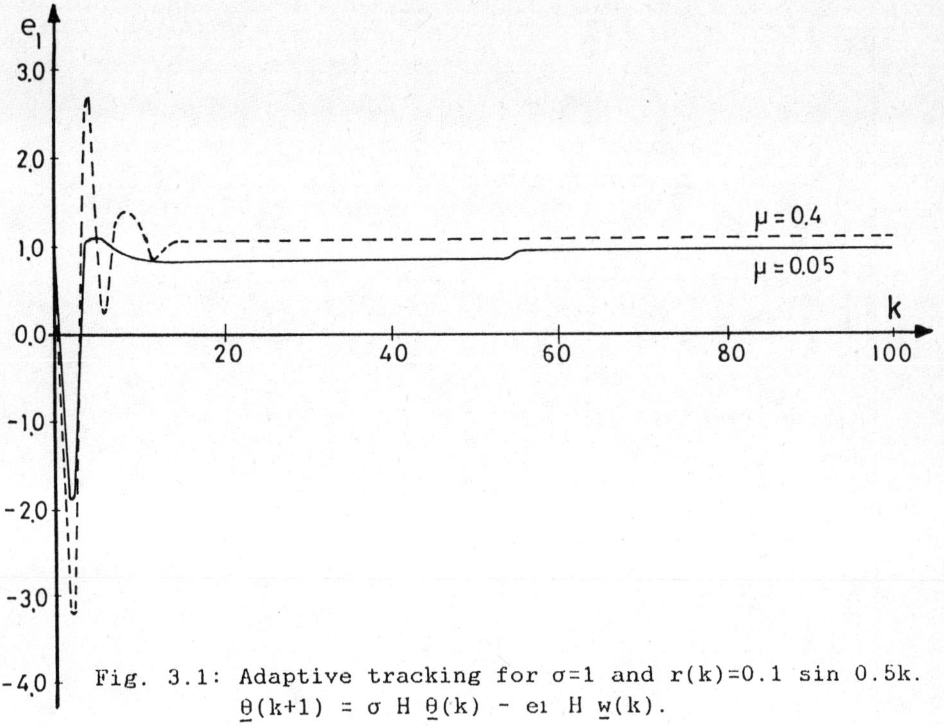

Fig. 3.1: Adaptive tracking for σ=1 and r(k)=0.1 sin 0.5k.
$\underline{\theta}(k+1) = \sigma H \underline{\theta}(k) - e_1 H \underline{w}(k)$.

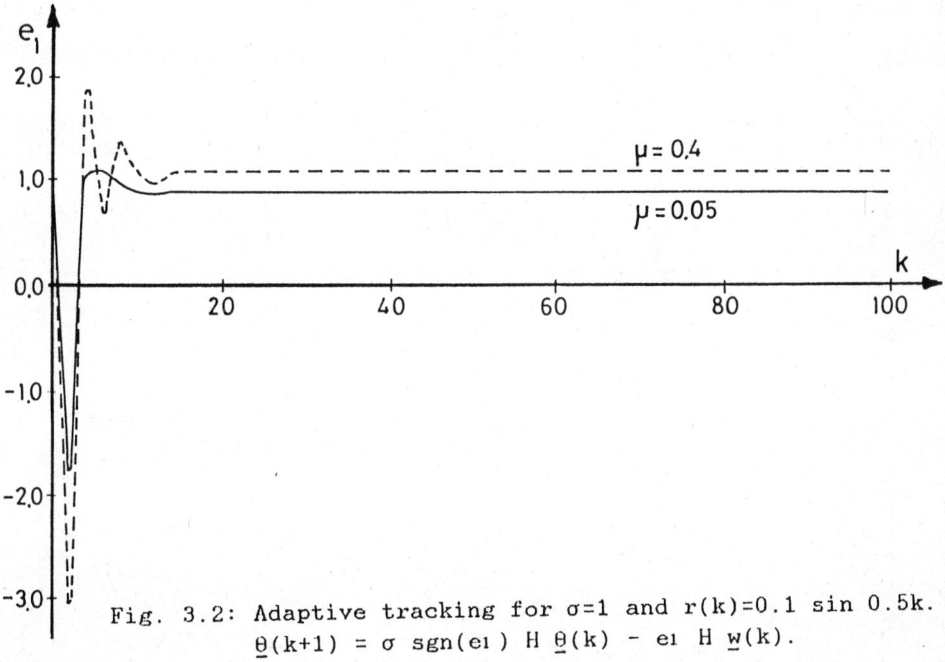

Fig. 3.2: Adaptive tracking for σ=1 and r(k)=0.1 sin 0.5k.
$\underline{\theta}(k+1) = \sigma\, sgn(e_1)\, H \underline{\theta}(k) - e_1 H \underline{w}(k)$.

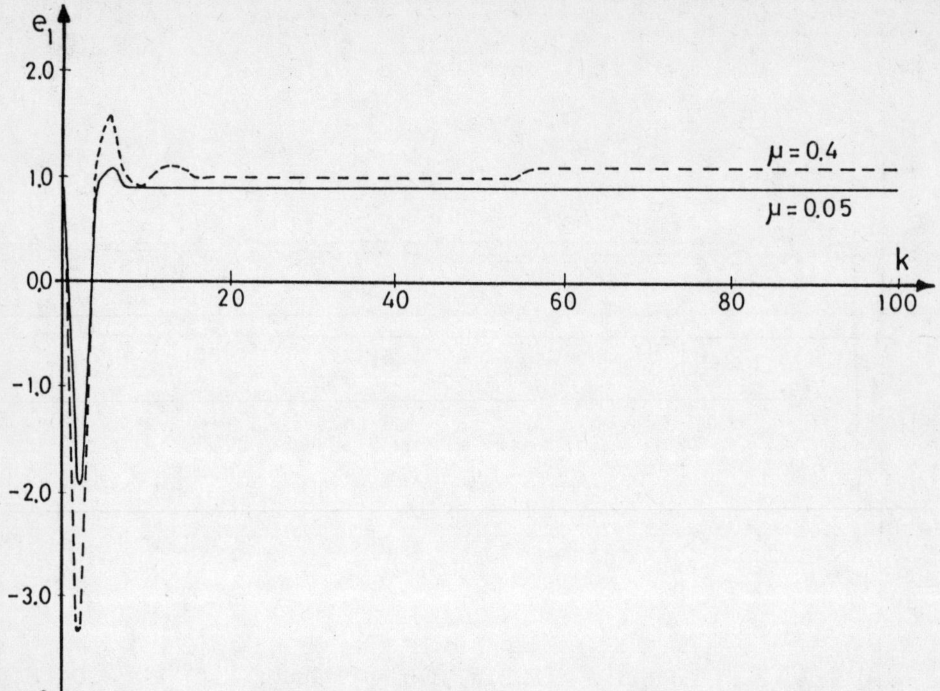

Fig. 3.3: Adaptive tracking for σ=1 and r(k)=0.1 sin 0.5k.
$\underline{\theta}(k+1) = \sigma \, sat(e_1) \, H \, \underline{\theta}(k) - e_1 \, H \, \underline{w}(k)$.

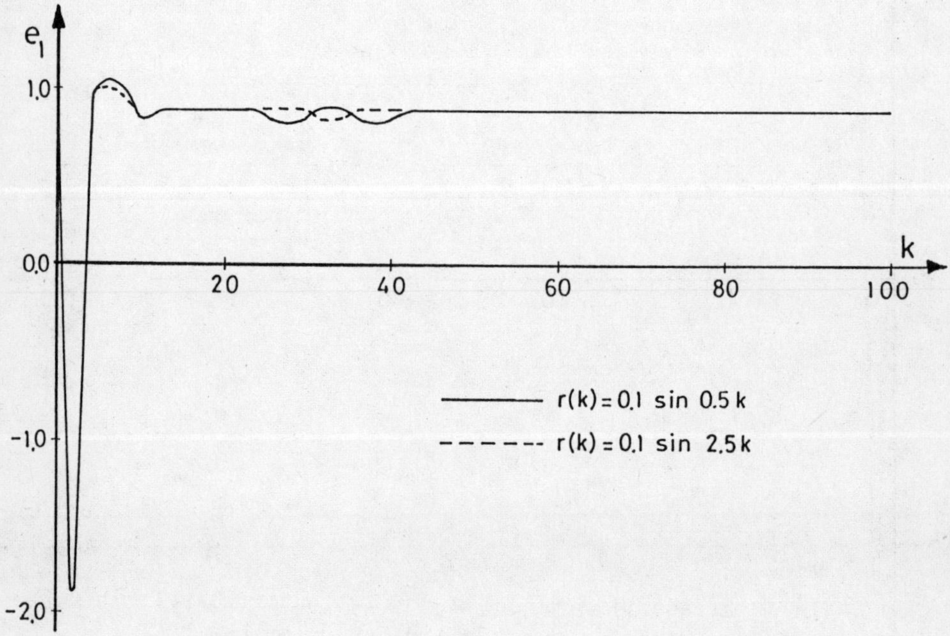

Fig. 3.4: Adaptive tracking for σ=0.2 and μ=0.05.
$\underline{\theta}(k+1) = \sigma \, H \, \underline{\theta}(k) - e_1 \, H \, \underline{w}(k)$.

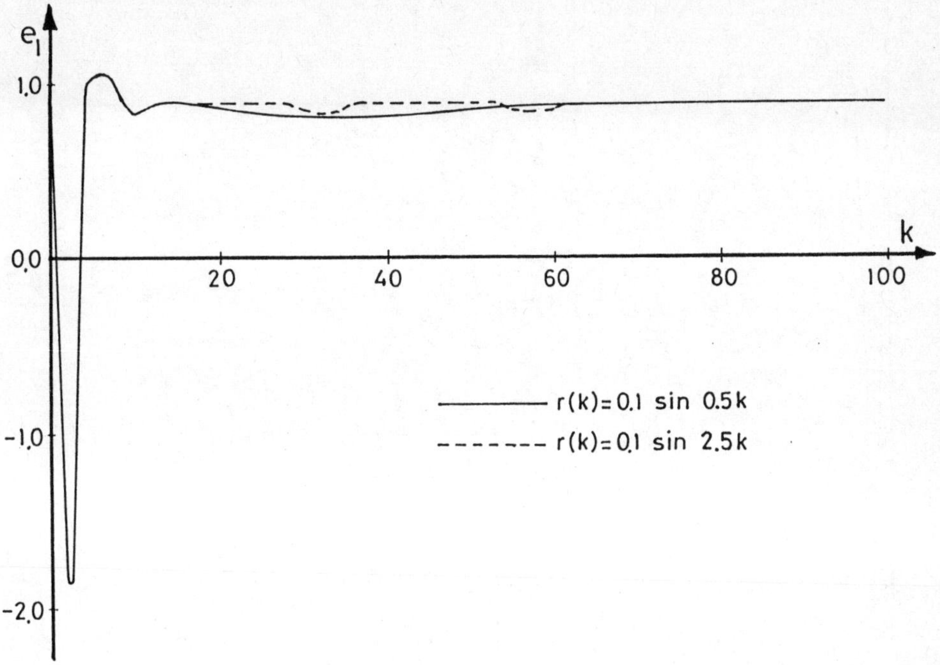

Fig. 3.5: Adaptive tracking for σ=0.2 and μ=0.05
$\underline{\theta}(k+1) = \sigma \, sgn(e_1) \, H \, \underline{\theta}(k) - e_1 \, H \, \underline{w}(k)$.

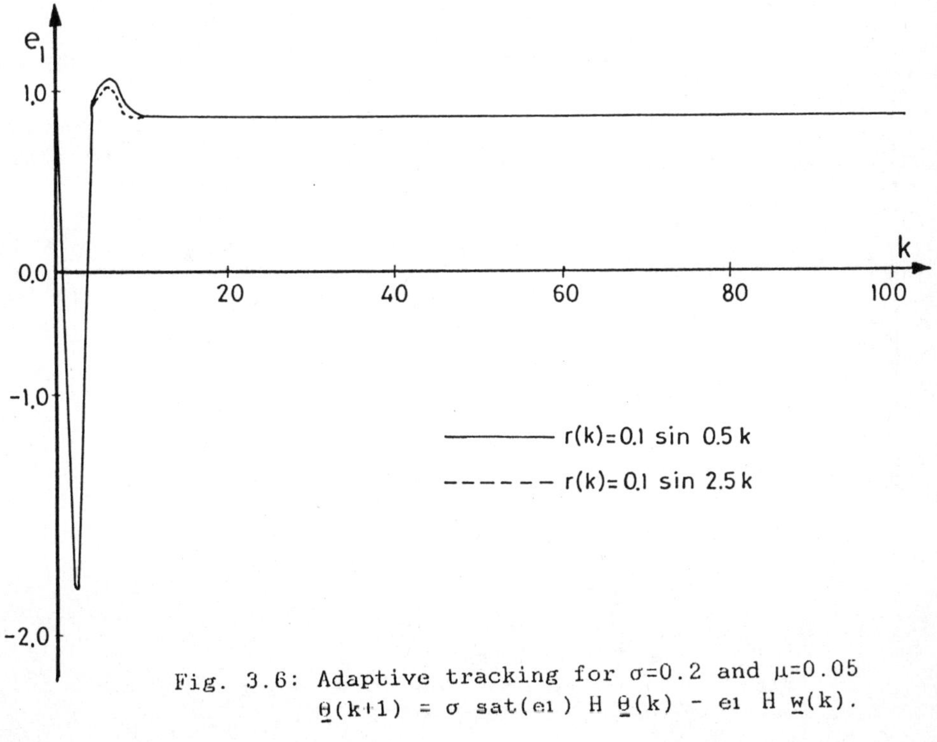

Fig. 3.6: Adaptive tracking for σ=0.2 and μ=0.05
$\underline{\theta}(k+1) = \sigma \, sat(e_1) \, H \, \underline{\theta}(k) - e_1 \, H \, \underline{w}(k)$.

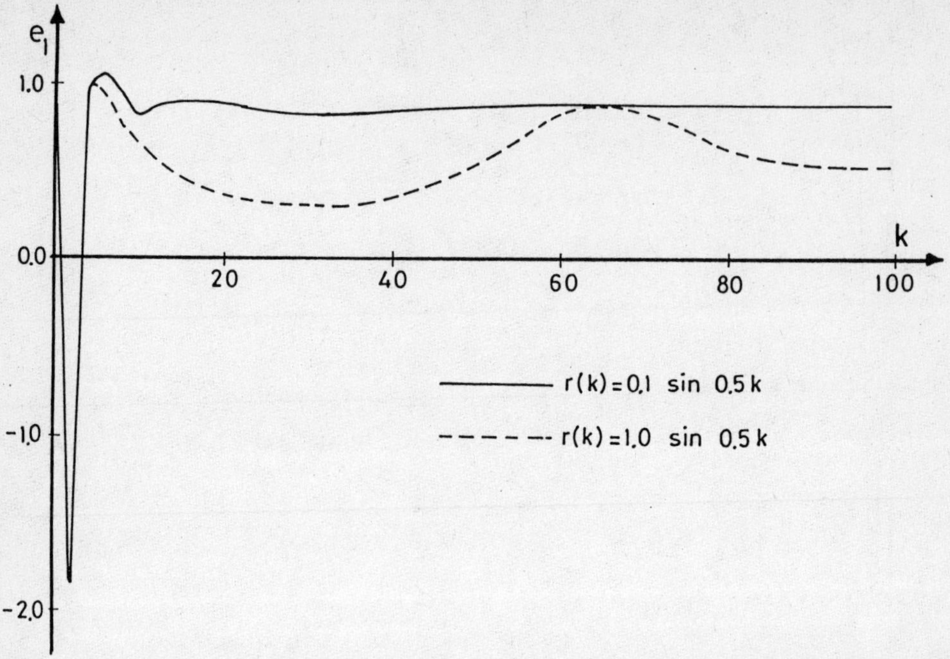

Fig. 3.7: Adaptive tracking for σ=0.2 and μ=0.05.
$\underline{\theta}(k+1) = \sigma H \underline{\theta}(k) - e_1 H \underline{w}(k)$.

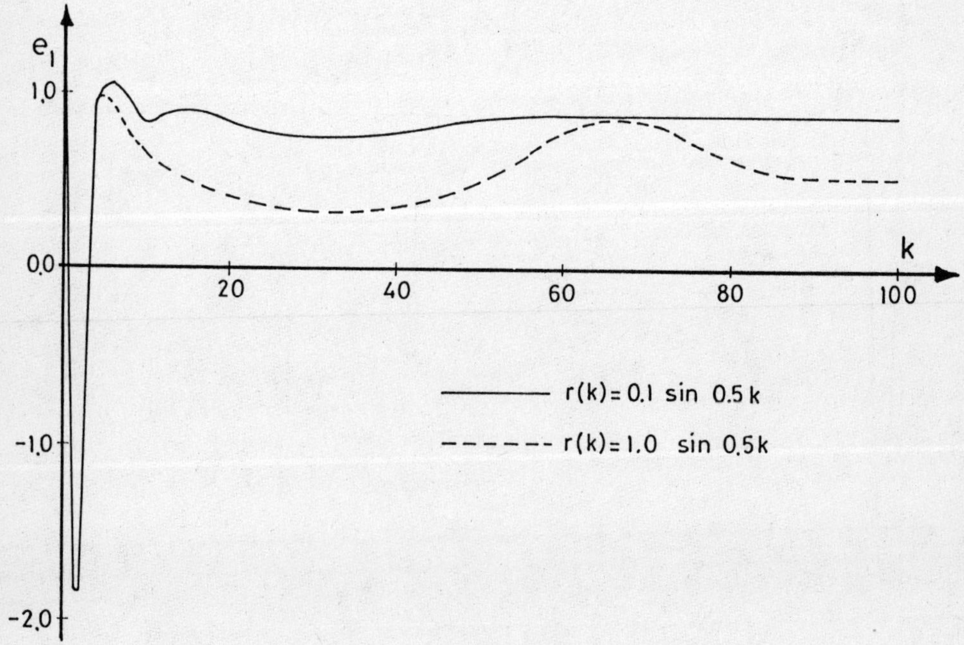

Fig. 3.8: Adaptive tracking for σ=0.2 and μ=0.05
$\underline{\theta}(k+1) = \sigma \, sgn(e_1) H \underline{\theta}(k) - e_1 H \underline{w}(k)$.

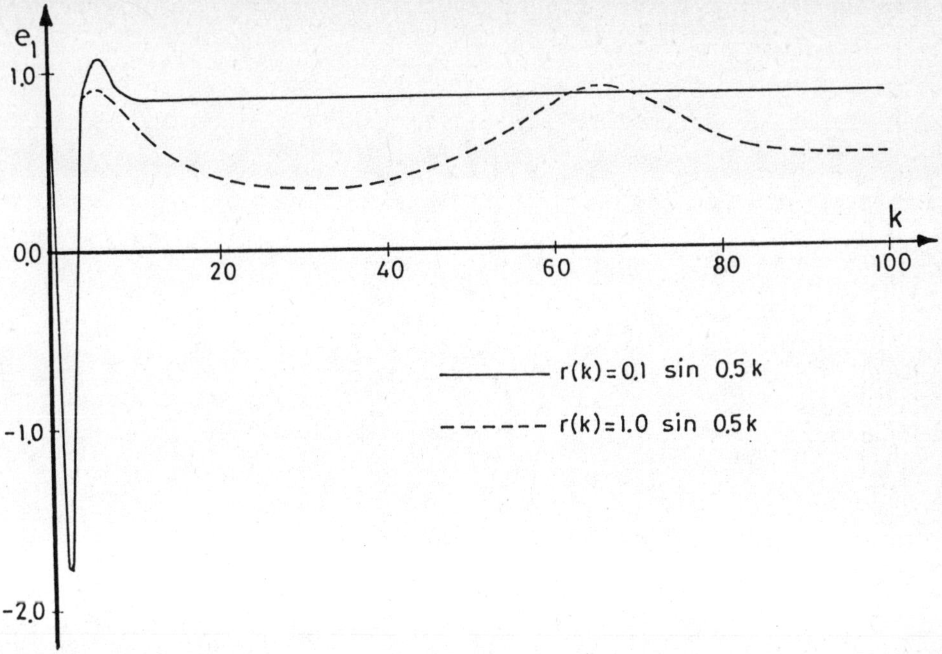

Fig. 3.9: Adaptive tracking for σ=0.2 and μ=0.05
$\underline{\theta}(k+1) = \sigma\ \mathrm{sat}(e_1)\ H\ \underline{\theta}(k) - e_1\ H\ \underline{w}(k)$.

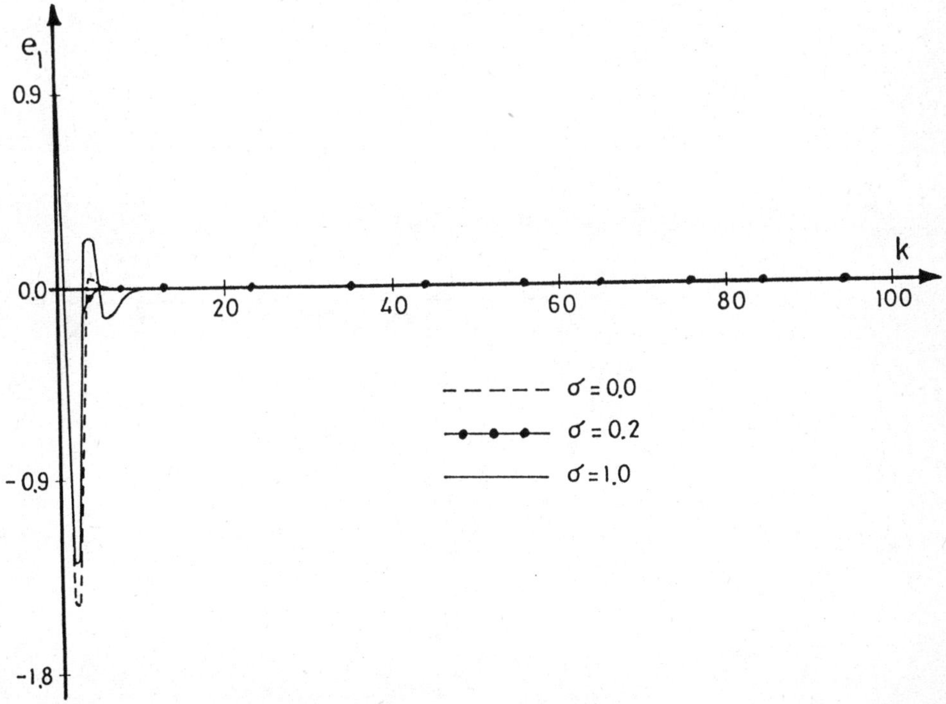

Fig. 3.10: Adaptive regulation for μ=0 without disturbance.
$\underline{\theta}(k+1) = \sigma\ H\ \theta(k) - e_1\ H\ \underline{w}(k)$.

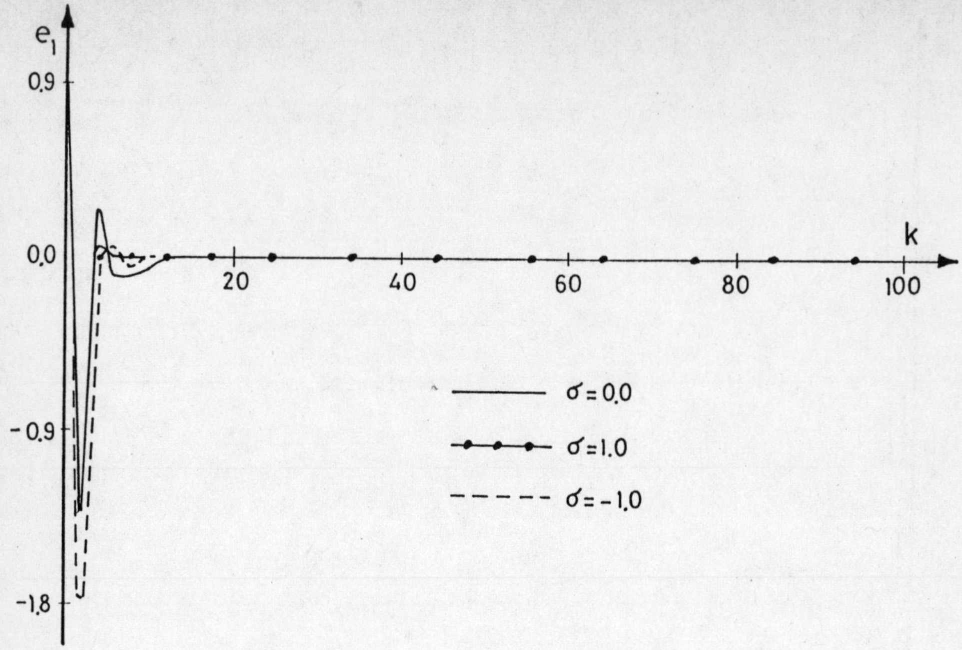

Fig. 3.11: Adaptive regulation for $\mu=0$ without disturbance.
$\underline{\theta}(k+1) = \sigma\ \mathrm{sgn}(e_1)\ H\ \underline{\theta}(k) - e_1\ H\ \underline{w}(k)$.

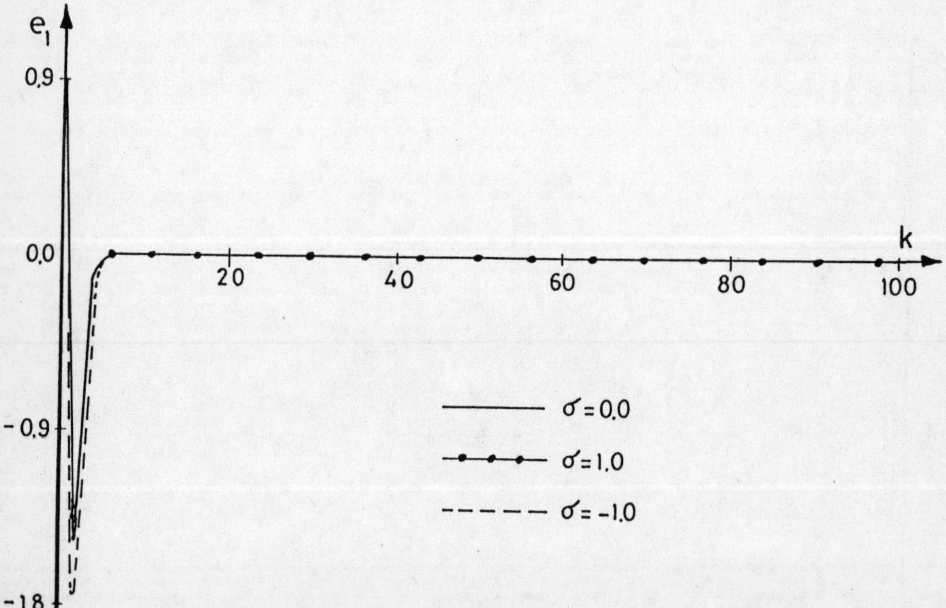

Fig. 3.12: Adaptive regulation for $\mu=0$ without disturbance.
$\underline{\theta}(k+1) = \sigma\ \mathrm{sat}(e_1)\ H\ \underline{\theta}(k) - e_1\ H\ \underline{w}(k)$.

$\sigma=0.2$, $r(k)=0.1 \sin .5k$ and two values of $\mu(0.05, 0.4)$. It has been observed that the error is bounded for the three adaptation schemes. Keeping the same conditions as before but changing the frequency of the input reference from 0.5 to 2.5, we can still achieve similar results as shown in Figs. 3.4-3.6. Increasing the magnitude of the reference input from 0.1 to 1.0 results in oscillatory but bounded response as depicted in Figs. 3.7-3.9. Finally, the simulation results of the adaptive regulation ($r(k)=0$, $\underline{X}_n(k)=\underline{0}$) and without disturbance with different values of σ for the adaptation laws (3.22), (3.51) and (3.55) respectively are plotted and compared with $\sigma=0$ in Figs. 3.10-3.12. Adaptive regulation is obtained for all adaptation laws suggested in this Chapter. We thus conclude that adaptive regulation and tracking can be guaranteed by appropriately changing the biasing term and/or the mode-separation ratio.

3.6 Conclusions

In this Chapter, we analyze reduced-order adaptive control schemes for linear discrete plants including bounded disturbances and/or unmodeled dynamics. Three adaptation mechanisms are presented and compared. It can be concluded that for uniformly bounded input, reduced-order controllers yield stable performance but within a conservative region of attraction. This region contains all signals which will converge to target sets around the equilibrium for perfect adaptive tracking. The size of the target sets depend on the adaptation gain, the mode separation ratio, the adaptation mechanism, the external bounded disturbance and finally the magnitude and periodicity of the reference input. Simulation results indicate that the design parameters can be adjusted to achieve good performance, under various operating conditions.

CHAPTER 4

DETERMINISTIC CONTROL OF DISCRETE SYSTEMS

4.1 Introduction

Stabilization of discrete systems with unknown bounded parameters and/or parasitics elements (henceforth termed uncertain systems) is a problem of paramount importance in process control engineering [1] and computer-control applications [2]. Several approaches have been developed to characterize the uncertainty and subsequently deal with different aspects of the cited problem [1,2]. In Chapter 1, an overview of adaptive control techniques has been presented. Stability properties of reduced-order adaptive systems for singulary-perturbed plants with bounded fast parasitics and/or external bounded disturbances, has been investigated in Chapters 2 and 3. Here, we are interested in a broad class of those systems in which the values of their uncertain parameters are only known to belong to given compact bounding intervals.

In the context of continuous-time, uncertain dynamical systems, a number of new approaches have been proposed for stabilizing such systems (e.g. see for example [38-42] and the references cited therein). However, as indicated in [68-70], such results cannot readily extended to the discrete case by direct discretization of the pertinent equations. Moreover, recent advances of microprocessors and related digital computer technology naturally favor the use of discrete systems in design and implementation. Consequently, the problem of stabilizing uncertain discrete systems is of interest in its own right.

In the present work, we concentrate on studying the

regulation problem for a class of linear time-invariant, discrete systems containing uncertain elements and/or additive-type uncertainty. The values of these uncertain parameters are unknown but bounded. That is, the values of these uncertainties are known to be contained within given compact bounding sets [68-70]. It can be established that when matching conditions hold, a two-part feedback control can be designed : a linear part to position the eigenvalues within the unit disk and a nonlinear (switching) part to ensure the uniform asymptotic stability ,or uniform ultimate boundedness behaviour of the uncertain dynamical system at hand.

In the sequel, we denote $\lambda_M(.)$ [$\lambda_m(.)$] as the operation of taking maximum [minimum] eigenvalue of (.),respectively. Given a vector $\underline{x}\in R^n$, We take $\|\underline{x}\| = (\underline{x}^T\underline{x})^{1/2}$. This induces the matrix norm $\|\phi\|=[\lambda_M(\phi^T\phi)]^{1/2}$. $\mathcal{D}^c$ denotes the complement of the set $\mathcal{D}$ and $\mathcal{Z}$ is set of all integers.

4.2 Problem Formulation

Consider a dynamical system that has an additive-type uncertainty shown in Fig. 4.1 and described by the following difference equation:

$$\underline{x}(k+1)=[A+\Delta A(\underline{r}(k))]\underline{x}(k) + [B+\Delta B(\underline{s}(k))]\underline{u}(k) + C\underline{v}(k) \tag{4.1a}$$

$$\underline{x}(k_0) = \underline{x}_0 \tag{4.1b}$$

where $\underline{x}\in R^n$ is the state, $\underline{u}\in R^m$ is the control, $\underline{v}\in R^l$ is the disturbance, A,B,C are prescribed constant matrices with appropriate dimensions. The uncertain parameters are $\underline{r}\in R^p$, $\underline{s}\in R^q$, and $\underline{v}\in R^l$. The system matrix uncertainty $\Delta A(\underline{r})$ and the input matrix uncertainty $\Delta B(\underline{s})$ depend on parameters $\underline{r}$ and $\underline{s}$ respectively. The term C $\underline{v}(k)$ accounts for input uncertainty

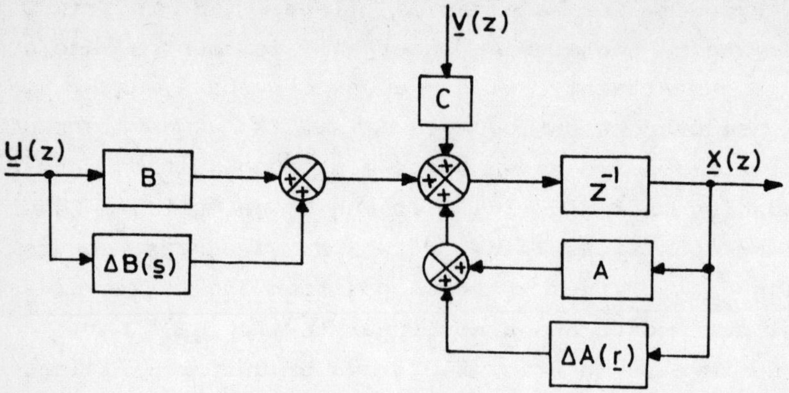

Figure 4.1: Additive-type Uncertain Discrete System

In the sequel, we assume that the following assumptions hold:

(A1) The entries of $\Delta A(.)$ and $\Delta B(.)$ are continuous on R^p and R^q respectively.

(A2) Uncertainty parameters $\underline{r}(.):R \longrightarrow \mathcal{R}$, $\underline{s}(.):R \longrightarrow \mathcal{S}$, and $\underline{v}(.):R \longrightarrow \mathcal{V}$ are Lebesgue measurable in compact bounding sets $\mathcal{R}$, $\mathcal{S}$ and $\mathcal{V}$ respectively, where

$$\mathcal{R} = \{ \underline{r} \in R^p : \check{r}_i < r_i < \hat{r}_i \ ; \ \check{r}_i, \hat{r}_i \quad \text{given constants} \}$$

$$\mathcal{S} = \{ \underline{s} \in R^q : \check{s}_i < s_i < \hat{s}_i \ ; \ \check{s}_i, \hat{s}_i \quad \text{given constants} \}$$

$$\mathcal{V} = \{ \underline{v} \in R^l : \check{v}_i < v_i < \hat{v}_i \ ; \ \check{v}_i, \hat{v}_i \quad \text{given constants} \}$$

(A3) {A,B} is a stabilizable pair; that is, there exists a

constant (mxn) matrix G, such that the eigenvalues $\lambda(\tilde{A}) = \lambda(A+BG)$ have moduli strictly less than unity.

Our objective is to design a nonlinear feedback control strategy in order to stabilize a class of linear discrete systems perturbed with uncertainties in the framework described above.

4.3 Guaranteed Asymptotic stability

For generality, we assume that the system matrix A is unstable. Now, given a matrix G satisfying assumption (A3), consider the class of feedback controls :

$$\underline{u}(k) = G\,\underline{x}(k) + \underline{g}(\underline{x}) \qquad \forall\ \underline{x} \in R^n \tag{4.2}$$

where $\underline{g}(.) : R_n \longrightarrow R^m$ is given by

$$g_i(\underline{x}) = \begin{cases} -\sigma_i(\underline{x})\ \mathrm{sgn}\ (\underline{b}_i^T\ P\ A\ \underline{x}) & \forall\ \underline{x} \notin N_i \\ \in \{\, \gamma_i \in R : |\gamma_i| \le \sigma_i(\underline{x})\} & \forall\ \underline{x} \in N_i \end{cases} \tag{4.3}$$

$$i \in \{\, 1,2,3,\ldots.,m\}$$

where the functions $\sigma_i(.) : R^n \longrightarrow R$ are non-negative functions defined later and $P > 0$ is a solution of Lyapunov equation

$$\tilde{A}^T P \tilde{A} - P = -Q \quad ; \qquad Q > 0 \tag{4.4}$$

vectors $\underline{b}_i$ is the ith column of matrix B, and

$$N_i = \{\underline{x} \in R^n : \underline{b}_i^T\ P\ A\ \underline{x} = 0\} \tag{4.5}$$

We are now in a position to state the so-called matching conditions [12] :

(A4) There exist matrix functions (of appropriate dimensions) D(.) and E(.) whose entries are continuous on R^p and R^q respectively such that:

$$\Delta A(\underline{r}) = BD(\underline{r}) \tag{4.6a}$$

$$\Delta B(s) = BE(\underline{s}) \tag{4.6b}$$

(A5) There exists a constant matrix function F such that:

$$C = BF \tag{4.6c}$$

With feedback control (4.2) and subject to assumptions (A4),(A5), the system (4.1) becomes :

$$\underline{x}(k+1) = A\,\underline{x}(k) + B\,\underline{g}(\underline{x}(k)) + B\,\underline{\Phi}(\underline{x}(k),k) \tag{4.7a}$$

$$\underline{x}(k_0) = \underline{x}_0 \tag{4.7b}$$

where

$$\underline{\Phi}(\underline{x}(k),k)=D(\underline{r}(k))\underline{x}(k)+E(\underline{s}(k))\underline{g}(\underline{x})+F\underline{v}(k)+E(\underline{s}(k))G\underline{x}(k) \tag{4.8}$$

Hence,

$$|\underline{\Phi}_i(\underline{x},k)| \le \max_{\underline{r}\in\mathcal{R}} |\underline{d}_i(r)\,\underline{x}| + \sum_{j=1}^{m} \max_{\underline{s}\in\mathcal{S}} |e_{ij}(\underline{s})|\,\sigma_i(\underline{x}) + \max_{\underline{v}\in\mathcal{V}} |\underline{f}_i\,\underline{v}| + \max_{\underline{s}\in\mathcal{S}} |\underline{e}_i(\underline{s})\,G\,\underline{x}| \tag{4.9}$$

where $\underline{e}_i$, $\underline{d}_i$ and $\underline{f}_i$ are the ith rows of matrices E, D and F respectively, and e_{ij} are the elements of the matrix E.

Next, we define

$$\Gamma = \begin{bmatrix} (1-\underset{\underline{s}\in\mathcal{S}}{\mathrm{Max}}\,|e_{11}(\underline{s})|) & -\underset{\underline{s}\in\mathcal{S}}{\mathrm{Max}}\,|e_{12}(\underline{s})| & -\underset{\underline{s}\in\mathcal{S}}{\mathrm{Max}}\,|e_{1m}(\underline{s})| \\ -\underset{\underline{s}\in\mathcal{S}}{\mathrm{Max}}\,|e_{21}(\underline{s})| & (1-\underset{\underline{s}\in\mathcal{S}}{\mathrm{Max}}\,|e_{12}(\underline{s})|) & -\underset{\underline{s}\in\mathcal{S}}{\mathrm{Max}}\,|e_{1m}(\underline{s})| \\ & & \\ -\underset{\underline{s}\in\mathcal{S}}{\mathrm{Max}}\,|e_{m1}(\underline{s})| & -\underset{\underline{s}\in\mathcal{S}}{\mathrm{Max}}\,|e_{m2}(\underline{s})| & (1-\underset{\underline{s}\in\mathcal{S}}{\mathrm{Max}}\,|e_{mm}(\underline{s})|) \end{bmatrix}$$

$$\underline{\alpha}(\underline{x}) = \begin{bmatrix} \alpha_1(\underline{x}) \\ \alpha_2(\underline{x}) \\ \\ \alpha_m(\underline{x}) \end{bmatrix}, \quad \underline{\beta}(\underline{x}) = \begin{bmatrix} \underset{\underline{r}\in\mathcal{R}}{\mathrm{Max}}\,|\underline{d}_1(\underline{r})\underline{x}| + \underset{\underline{v}\in\mathcal{V}}{\mathrm{Max}}\,|\underline{f}_1\underline{v}| + \underset{\underline{s}\in\mathcal{S}}{\mathrm{Max}}\,|\underline{e}_1(\underline{s})G\underline{x}| \\ \\ \underset{\underline{r}\in\mathcal{R}}{\mathrm{Max}}\,|\underline{d}_m(\underline{r})\underline{x}| + \underset{\underline{v}\in\mathcal{V}}{\mathrm{Max}}\,|\underline{f}_m\underline{v}| + \underset{\underline{s}\in\mathcal{S}}{\mathrm{Max}}\,|\underline{e}_m(\underline{s})G\underline{x}| \end{bmatrix} \tag{4.10}$$

and finally assume that :

(A6) The matrix Γ is nonsingular.

(A7) For all $\underline{x} \in R^n$, the components of $\underline{\alpha}(\underline{x})$ are non-negative, where $\underline{\alpha}(\underline{x})$ is the solution of

$$\Gamma\, \underline{\alpha}(\underline{x}) = \underline{\beta}(\underline{x}) \tag{4.11}$$

Now, in view of assumptions (A6),(A7) we may define :

$$\sigma_i(\underline{x}) = \alpha_i(\underline{x}) \quad , \quad i = 1,2,3,\ldots,m \tag{4.12}$$

so that, by (4.9), for $i \in \{1,2,3,\ldots,m\}$,

$$|\Phi_i(\underline{x},k)| \leq \sigma_i(\underline{x}) \qquad \forall \quad (\underline{x},k) \in R^n \times \mathcal{Z} \tag{4.13}$$

It is easy to see that the entries of $\underline{d}_i(.)$ and $\underline{e}_i(.)$ are continuous, $\mathcal{R}$ and $\mathcal{S}$ are compact, $\underline{\alpha}(.)$ and hence $\sigma(.)$ are continuous.

Theorem 4.3.1

Consider system (4.1) with control (4.2). If assumptions (A1-A7) are met, then for any admissible uncertainties $\underline{r}(.)$, $\underline{s}(.)$, and $\underline{v}(.)$, and any initial condition $(\underline{x}_0, k_0)$, the state $\underline{x} = \underline{0}$ is uniformly asymptotically stable in the sense of Lyapunov.

Proof

It is clear from (4.3) that $\underline{g}(.)$ is discontinuous and hence multivalued for $(\underline{x},k) \in N_i \times \mathcal{Z}$. Consequently, we regard (4.7) as a generalized dynamical system [10]. In particular, one must show that :

(a) Given any initial condition $(\underline{x}_0, k_0) \in R^n \times \mathcal{Z}$, there is at least one solution of (4.7).

(b) There is a candidate Lyapunov function with strictly negative difference along every nontrivial solution.

(c) $\underline{x}(k) = \underline{0}$ is a possible solution.

To prove a,b, and c, we proceed as follows :

(a) Consider the set-valued function $F(.):D \subset R^n \times \mathcal{Z} \longrightarrow$ all non-empty sets of R^n, where

$$F(\underline{x},k) \triangleq \{\underline{z}\in R^n : \underline{z} = [A + \Delta A(\underline{r}(k))]\underline{x}(k) + [B + \Delta B(\underline{s}(k))]G\underline{x}(k)$$

$$+ \sum_{i=1}^{q} [\underline{b}_i^T + \Delta\underline{b}_i^T(\underline{s}(k))]y_i(k) - \sum_{i=q+1}^{m} [\underline{b}_i + \Delta\underline{b}_i^T(\underline{s}(k))] \bullet$$

$$\bullet \sigma_i(\underline{x})\ \mathrm{sgn}(\underline{b}_i^T PA\underline{x}) + C\underline{v}(k),\ |y_i(k)| \le \sigma_i(\underline{x}) \qquad (4.14)$$

for all $(\underline{x},k) \in [\bigcap_{i=1}^{q} (N_i \times \mathcal{Z})] \cap D$,

where $\Delta\underline{b}_i$ is the ith column of the matrix ΔB.

Then, $\underline{x}(.):[k_0,k_1] \longrightarrow R^n$ is a solution of (4.7) if it satisfies

$$\underline{x}(k_0) = \underline{x}_0 \qquad \text{and} \qquad \underline{x}(k+1) \in F(\underline{x}(k),k) \qquad (4.15)$$

The existence of such a solution is assured if the following conditions are met for all $(\underline{x},k) \in D$:

(i) $F(\underline{x},k)$ is compact.
(ii) $F(\underline{x},k)$ is convex.
(iii) $F(.,k)$ is upper semicontinuous.
(iv) There is a measurable function $h(\underline{x},.) : [k_0,k_1] \longrightarrow R^n$ such that $h(\underline{x},k) \in F(\underline{x},k)$.
(v) There exists a function $g(.) : [k_0,k_1] \longrightarrow R$ such that $\|\underline{z}\| \le g(k) \quad \forall\ \underline{z} \in F(\underline{x},k)$.

To show (i),(ii), it suffices to note that $F(\underline{x},k)$ is either a compact or convex set.

To prove (iii), let

$$\underline{\theta}(\underline{x},k) = [A + \Delta A(\underline{r}(k))]\underline{x}(k) + [B + \Delta B(\underline{s}(k))]G\underline{x}(k) + C\underline{v}(k)$$

and consider $F(\bar{\underline{x}},\bar{k})$ at $(\bar{\underline{x}},\bar{k}) \notin N_i \times \mathcal{Z}$, $i=1,2,...,m$

and at $(\bar{\underline{x}},\bar{k}) \in \bigcap_{i=1}^{q\leq m} N_i \times \mathcal{B}$.

In the first case, condition (iii) is met since $F(.,\bar{k})$ is continuous. In the second case, $F(.,\bar{k})$ is upper-semicontinuous, [71], if the separation between $F(\underline{x},\bar{k})$ and $F(\bar{\underline{x}},\bar{k})$ is continuous for sufficiently small $\|\bar{\underline{x}} - \underline{x}\|$, $(\underline{x},\bar{k}) \notin N_j \times \mathcal{B}$ for $j \in \{1,2,....,q\}$.

Thus, suppose $(\underline{x},\bar{k}) \in \bigcap_{i=1}^{l\leq q} N_i \times \mathcal{B}$, then the separation :

$$d^*(F(\bar{\underline{x}},\bar{k}),F(\underline{x},\bar{k})) = \sup_{y_i} \operatorname{Inf}_{\bar{y}_i} \| \underline{\theta}(\bar{\underline{x}},\bar{k}) - \underline{\theta}(\underline{x},\bar{k}) +$$

$$+ \sum_{i=1}^{l} [\, \underline{b}_i^T + \Delta\underline{b}_i^T(\underline{s}(\bar{k}))]\,(\bar{y}_i - y_i) +$$

$$+ \sum_{i=l+1}^{q} [\underline{b}_i^T + \Delta\underline{b}_i^T(\underline{s}(k))](\bar{y}_i - \sigma_i(\underline{x})\operatorname{sgn}(\underline{b}_i^T PA\underline{x})] +$$

$$+ \sum_{i=q+1}^{m} [\underline{b}_i^T + \Delta\underline{b}_i^T(\underline{s}(\bar{k}))].[- \sigma_i(\bar{\underline{x}})\ \operatorname{sgn}(\underline{b}_i^T PA\bar{\underline{x}}) +$$

$$+ \sigma_i(\underline{x})\ \operatorname{sgn}(\underline{b}_i^T PA\underline{x})]\|$$

$$\therefore d^* \leq \| \underline{\theta}(\bar{\underline{x}},\bar{k})-\underline{\theta}(\underline{x},\bar{k})\| + \sup_{y_i} \inf_{\bar{y}_i} \sum_{i=1}^{l} \| [\underline{b}_i^T + \Delta\underline{b}_i^T(\underline{s}(\bar{k}))](\bar{y}_i - y_i)\| +$$

$$+ \operatorname*{Inf}_{\bar{y}_i} \sum_{i=l+1}^{q} \| [\underline{b}_i^T + \Delta\underline{b}_i^T (\underline{s}(\bar{k}))](\bar{y}_i + \sigma_i(\underline{x})\,\mathrm{sgn}(\underline{b}_i^T PA\underline{x}))\| +$$

$$+ \sum_{i=q+1}^{m} \| [\underline{b}_i^T + \Delta\underline{b}_i^T (\underline{s}(\bar{k}))].[-\sigma_i(\bar{\underline{x}})\,\mathrm{sgn}(\underline{b}_i^T PA\bar{\underline{x}}) + \sigma_i(\underline{x})\,\mathrm{sgn}(\underline{b}_i^T PA\underline{x})]\| \tag{4.16}$$

For sufficiently small $\|\bar{\underline{x}} - \underline{x}\|$,

$$\mathrm{sgn}\,(\underline{b}_i^T PA\bar{\underline{x}}) = \mathrm{sgn}\,(\underline{b}_i^T PA\underline{x}) \quad \text{for} \quad i \in \{q+1, q+2, \ldots, m\}$$

Also, we may choose,

$$\bar{y}_i = -\sigma_i(\bar{\underline{x}})\ \mathrm{sgn}\,(\underline{b}_i^T PA\underline{x}) \quad \text{for} \quad i \in \{l+1, l+2, \ldots q\}$$

In the second term of (4.16), we must consider two cases :

C1. $\sigma_i(\bar{\underline{x}}) \geq \sigma_i(\underline{x})$

In this case, no matter what y_i is, we must have $\bar{y}_i = y_i$ to vanish the second term of (4.16).

C2. $\sigma_i(\bar{\underline{x}}) < \sigma_i(\underline{x})$

Then, we must have $|y_i| = \sigma_i(\underline{x})$.
But since $|y_i| > \sigma_i(\bar{\underline{x}})$ make the second term of (4.16) vanish again. But then $y_i\ \bar{y}_i = [y_i\ /\ |y_i|]\ \sigma_i(\bar{\underline{x}}(k))$.

Combining all the above, (4.16) can be written as :

$$d^*(F(\bar{\underline{x}},\bar{k}), F(\underline{x},\bar{k})) \leq \|\underline{\theta}(\bar{\underline{x}},\bar{k}) - \underline{\theta}(\underline{x},\bar{k})\| +$$

$$+ \sum_{i=1}^{m} \| [\underline{b}_i^T + \Delta\underline{b}_i^T (\underline{s}(\bar{k}))].[\sigma_i(\bar{\underline{x}}) - \sigma_i(\underline{x})]\| \tag{4.17}$$

where the terms for $i \in \{1,2,\ldots,l\}$ are vanished if $\sigma_i(\bar{\underline{x}}) \geq \sigma_i(\underline{x})$.

Note that the right hand side of (4.17) is the sum of terms, each of which is the norm of the difference of functions whose entries are continuous of $\underline{x}$. This ensures the continuity of the separation.

Also, the validity of condition (iv) follows directly from assumptions (A1) and (A2).

Now, to prove (v), consider the set $D = D_0 \times [k_0, k_1]$ with D_0 being compact subset of R^n. As we shall show in proving (b), this suffices since no solution can leave a certain compact subset of R^n.

In view of definition (4.14),

$$\|\underline{z}\| \leq \|A + B\,G\|\,\tilde{\underline{x}} + \underset{\underline{r}\in\mathcal{R}}{\mathrm{Max}}\,\|\Delta A(\underline{r}(k))\|\,\tilde{\underline{x}} + \underset{\underline{s}\in\mathcal{S}}{\mathrm{Max}}\,\|\Delta B(\underline{s}(k))\,G\|\,\tilde{\underline{x}} +$$

$$+ \sum_{i=1}^{m} [\|\underline{b}_i\| + \underset{\underline{s}\in\mathcal{S}}{\mathrm{Max}}\,\|\Delta \underline{b}_i(\underline{s}(k))\|\,\tilde{\sigma}_i + \underset{\underline{v}\in\mathcal{V}}{\mathrm{Max}}\,\|C\,\underline{v}(k)\|$$

$$= \text{constant} < \infty \qquad \forall\ \underline{z} \in F(\underline{x}(k),k)$$

where $\tilde{\underline{x}} = \underset{\underline{x}\in D_0}{\mathrm{Max}}\,\|\underline{x}\|$, $\tilde{\sigma}_i = \underset{\underline{x}\in D_0}{\mathrm{Max}}\,\sigma_i(\underline{x})$

(b) To prove (b), consider a candidate Lyapunov function : $V(.) : R^n \longrightarrow R_+$ given by

$$V(\underline{x}(k)) = \underline{x}^T(k)\,P\,\underline{x}(k) \tag{4.18}$$

where $P = P^T > 0$ is a solution of (4.4).

Now, for all pairs $(\underline{x},k) \in \bigcap_{i=1}^{q\leq m} N_i \times \mathfrak{Z}$ and $\underline{x} \neq \underline{0}$, the Lyapunov forward difference corresponding to the resulting closed-loop system (4.7) and the Lyapunov function (4.18) is given by :

$$\Delta V(k) = V(\underline{x}(k+1)) - V(\underline{x}(k)) = \underline{x}^T(k+1)P\underline{x}(k+1) - \underline{x}^T(k)P\underline{x}(k)$$

$$= \underline{x}^T(k)[\tilde{A}^T P \tilde{A} - P]\underline{x}(k) + 2\, \underline{x}^T(k)\, \tilde{A}^T P B\, \underline{g}(\underline{x}(k)) +$$

$$+ 2\, \underline{x}^T(k)\, \tilde{A}^T P B\, \underline{\Phi}(\underline{x}(k),k) + \underline{g}^T(\underline{x}(k))\, B^T P B\, \underline{g}(\underline{x}(k)) +$$

$$+ 2\, \underline{g}^T(\underline{x}(k))B^T PB\underline{\Phi}(\underline{x}(k),k) + \underline{\Phi}^T(\underline{x}(k),k)B^T PB\underline{\Phi}(\underline{x}(k),k)$$

Dropping the suffices for simplicity and using (4.3), (4.4), and (4.13), we have :

$$\Delta V(k) = -\, \underline{x}^T Q\, \underline{x} + 2\, \underline{x}^T \tilde{A}^T PB\, [\underline{g}(\underline{x})+\underline{\Phi}(\underline{x},k)] +$$

$$+ \underline{g}^T(\underline{x})B^T PB\, [\underline{g}(\underline{x})+\underline{\Phi}(\underline{x},k)] + [\underline{g}(\underline{x})+\underline{\Phi}(\underline{x},k)]^T\, B^T PB\, \underline{\Phi}(\underline{x},k)$$

$$= -\underline{x}^T Q\underline{x} + 2 \sum_{i=q+1}^{m} \underline{b}_i^T P\tilde{A}\underline{x}\, [-\sigma_i(\underline{x})\, \mathrm{sgn}(\underline{b}_i^T PA\underline{x}) + \Phi_i(\underline{x},k)]$$

$$+ \underline{g}^T(\underline{x})B^T PB \sum_{i=q+1}^{m} [-\sigma_i(\underline{x})\, \mathrm{sgn}(\underline{b}_i^T PA\underline{x}) + \Phi_i(\underline{x},k)] +$$

$$+ \{\sum_{i=q+1}^{m} [-\sigma_i(\underline{x})\, \mathrm{sgn}(\underline{b}_i^T PA\underline{x}) + \Phi_i(\underline{x},k)]\}^T\, B^T PB\underline{\Phi}(\underline{x},k)$$

$$\leq -\, \underline{x}^T Q\, \underline{x} < 0 \qquad (4.19)$$

Here, we note also that, given $(\underline{x}_0, k_0)$, every solution $\underline{x}(.)$ with $\underline{x}(k_0) = \underline{x}_0$ is such that

$$\underline{x}(k) \in X_o = \{\underline{x} \in R^n : \underline{x}^T P \underline{x} \le \underline{x}_o^T P \underline{x}_o\} \quad \forall \; k \ge k_o$$

Thus, we may take $D_o = X_o$.

(c) Finally, to prove (c), it is suffices to note that $\{\underline{0}\} \in F(\underline{0},k)$ so that $\underline{x}(k) = \underline{0}$ is possible solution of (4.7).

This concludes the proof of Theorem 4.3.1. ▪▪▪

4.4 Example and Discussion

To illustrate the results of the above Theorem, we consider a second order system shown in Fig. 4.2 with $r_1 \in [-0.5,1]$, $r_2 \in [-0.5,1]$, $s \in [-0.25,0.5]$ and $v \in [-1,1]$.

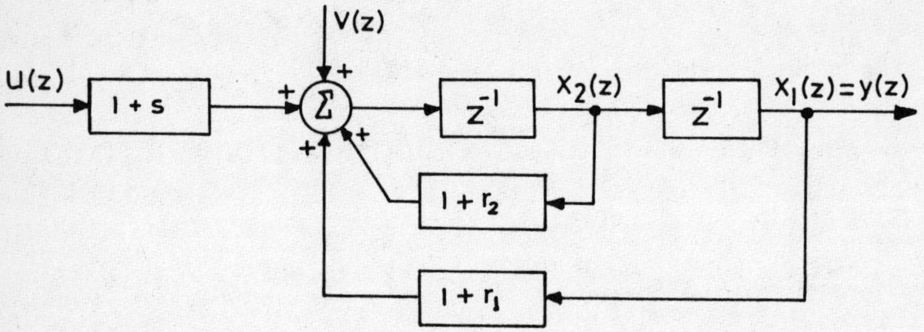

Figure 4.2: Uncertain Second order Linear System.

The above system can be cast in the form (4.1) with

$$A = \begin{bmatrix} 0 & 1 \\ 1 & 1 \end{bmatrix}, \quad \Delta A(\underline{r}) = \begin{bmatrix} 0 & 0 \\ r_1 & r_2 \end{bmatrix}, \quad B = \begin{bmatrix} 0 \\ 1 \end{bmatrix}, \quad \Delta B(\underline{s}) = \begin{bmatrix} 0 \\ s \end{bmatrix}, \quad C = \begin{bmatrix} 0 \\ 1 \end{bmatrix}$$

The eigenvalues of the system matrix A, are $1/2 \pm \sqrt{5}/2$, i.e. A is unstable and hence we require initially linear feedback; for instance, we may take $G = [-9/8 \quad -1/4]$ so that

$$\tilde{A} \triangleq A + B\,G = \begin{bmatrix} 0 & 1 \\ -1/8 & 3/4 \end{bmatrix}$$

Hence, with $Q = I_2$, the solution of (4.4) is

$$P = \begin{bmatrix} 37/35 & -32/105 \\ -32/105 & 128/35 \end{bmatrix}$$

Moreover, the matching conditions are satisfied with $D(\underline{r}) = [r_1 \quad r_2]$, $E(s) = s$, $F = 1$, so that

$$\sigma(\underline{x}) = \max_{\underline{r}\in\mathcal{R}} |r_1 x_1 + r_2 x_2| + \max_{s\in\mathcal{S}} |s|\,\sigma(\underline{x}) + \max_{v\in\mathcal{V}} |v| + {}$$
$$+ \max_{s\in\mathcal{S}} |s(-9x_1/8 - x_2/4)|$$

i.e. $\sigma(\underline{x}) \le 2\,[\,1 + |x_1| + |x_2| + |9x_1/16 + x_2/8|$

$$g(\underline{x}) = -\sigma(\underline{x})\, B^T P A \underline{x} \,/\, \|B^T P A \underline{x}\|$$

$$= -\sigma(\underline{x}).[\mathrm{sgn}(128x_1/35 + 352x_2/105)]$$

The complete feedback control is given by :

$$u(k) = G\,\underline{x}(k) + g(\underline{x})$$

$$= [-9x_1/8 - x_2/4] - \sigma(\underline{x})\,\mathrm{sgn}(128x_1/35 + 352x_2/105)$$

The closed-loop uncertain system was simulated with the above controller structure and the resultant closed-loop state trajectories are plotted in Fig. 4.3. It can be concluded that the developed nonlinear control ensures uniform asymptotic behaviour of the origin for a wide class of bounded admissible uncertainties.

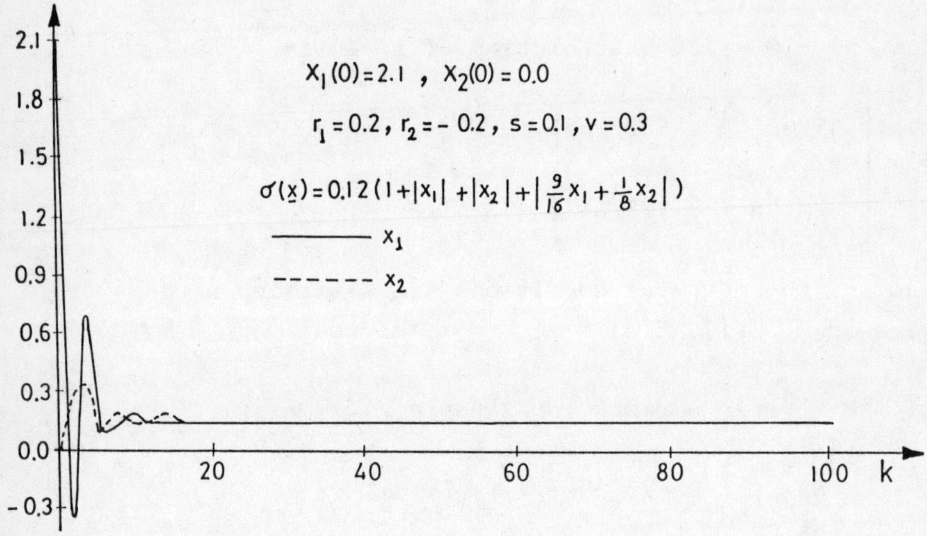

Fig. 4.3: Simulation Example

It is worth noting that the proposed control scheme typically requires switching of the control signal values on hyperplanes in the state space and one may argue that such control structure is difficult to implement. Furthermore, it may suffice in practice to assure that all solutions are ultimately bounded, no matter what the uncertainties are; that one may want to guarantee that every solution enters a neighborhood of the zero state in finite time and thereafter remains within that neighborhood. This can be accomplished by approximating the relay action with saturation one, i.e. the nonlinear (switching) part, $\underline{g}(\underline{x})$, in (4.2) is replaced by saturation control as shown in the next section.

4.5 Uniform Bounded Stabilization

In this section, our objective is to design a class of feedback controller structures that can guarantee uniform and ultimate boundedness for the system at hand irrespective of uncertainties and initial conditions.

Before going on the analysis, we have to recall some preliminary results of Lyapunov theory [18-20]

Theorem 4.5.1 (Boundedness of Solutions by Lyapunov's Direct Method)

For a system represented by

$$\underline{x}(k+1) = F\underline{x}(k) + \underline{f}(\underline{x}(k),k) \qquad (4.20)$$

where $\underline{x} \in R^n$ is the state, $F \in R^{n \times n}$ is the system matrix and $\underline{f}(.) \in R^n$ is in general nonlinear, time-varying, bounded input. Let $\mathcal{D}$ be a bounded neighborhood of the origin and let $\mathcal{D}^c$ be its complement (that is, $\mathcal{D}^c$ is the set of all points outside $\mathcal{D}$). Assume that $V(\underline{x})$ is a scalar function with its forward difference $\Delta V(\underline{x})$ satisfying the conditions:

(a) $V(\underline{x}) > 0 \quad \forall\, \underline{x} \in \mathcal{D}^c$

(b) $\Delta V(\underline{x}) \leq 0 \quad \forall\, \underline{x} \in \mathcal{D}^c$

(c) $V(\underline{x}) \longrightarrow \infty \quad \text{as } \|\underline{x}\| \longrightarrow \infty$

then each solution of (4.20) is bounded $\forall\, k \geq k_0$

Definition 4.5.1 (Uniform Boundedness)

Given a solution $\underline{x}(.)$: $[k_0, k_1] \longrightarrow R^n$; $\underline{x}(k_0) = \underline{x}_0$, of (4.20), then it is uniformly bounded if there exists a positive constant $\delta(\underline{x}_0)<\infty$, possibly dependent on $\underline{x}_0$ but not on k_0 such that $\|\underline{x}(k)\| \le \delta(\underline{x}_0) \quad \forall\ k\in[k_0, k_1]$

Definition 4.5.2 (Uniform Ultimate Boundedness)

Given a solution $\underline{x}(.):[k_0,\infty[\longrightarrow R^n$; $\underline{x}(k_0)=\underline{x}_0$, of (4.20), then it is uniformly ultimately bounded with respect to a set Ω if there exists a non-negative constant $\tilde{K}(\underline{x}_0,\Omega) < \infty$, possibly dependent on $\underline{x}_0$ and Ω but not on k_0 , such that $\underline{x}(k) \in \Omega \quad \forall\ k \ge k_0 + \tilde{K}(\underline{x}_0,\Omega)$

4.6 Measured State Feedback Control

For generality, we suppose that there may be measurement error in the state. Thus the measured state is given by :

$$\underline{y}(k) = \underline{x}(k) + \underline{w}(k) \tag{4.21}$$

where $\underline{w}(k) \in R^n$ is the measurement error and is assumed to be Lebesgue measurable in a compact bounding set $\mathcal{W}$, where

$$\mathcal{W} = \{\ \underline{w} \in R^l\ :\ \check{w}_i < w_i < \hat{w}_i \quad ;\ \check{w}_i\ ,\ \hat{w}_i \quad \text{given constants}\ \}$$

Again, we assume that the system matrix A is unstable, Now, given a matrix G satisfying assumption (A3), consider the class of feedback controls:

$$\underline{u}(k) = G\underline{y}(k) + \underline{g}(\underline{y}(k)) \qquad \forall\ \underline{y}\in R^n \tag{4.22}$$

where $\underline{g}(.):R^n \longrightarrow R^m$ is given by

$$\underline{g}(\underline{y}) = \begin{cases} -\sigma(\underline{y})B^T P\tilde{A}\underline{y}/\|B^T P\tilde{A}\underline{y}\| & \text{for } \|B^T P\tilde{A}\underline{y}\| > \epsilon \\ -\sigma(\underline{y})B^T P\tilde{A}\underline{y}\ /\ \epsilon & \text{for } \|B^T P\tilde{A}\underline{y}\| \le \epsilon \end{cases} \tag{4.23}$$

where ϵ is a prespecified positive constant (chosen by the designer), P is the solution of (4.4), and the function $\sigma(.)$: $R^n \longrightarrow R^m$ is non-negative function will be defined later.

With the application of nonlinear control (4.22),(4.23) and subject to assumptions (A1-A5), system (4.1) becomes

$$\underline{x}(k+1) = A\,\underline{x}(k) + B\,\underline{g}(\underline{y}(k)) + B\,\underline{\Phi}(\underline{x}(k),k) \tag{4.24a}$$

$$\underline{x}(k_0) = \underline{x}_0 \tag{4.24b}$$

where $\forall\ (\underline{x},k) \in R^n \times \mathcal{Z}$

$$\underline{\Phi}(\underline{x},k) = D(\underline{r}(k))\underline{x}(k) + G\underline{w}(k) + E(\underline{s}(k))G(\underline{x}(k)+\underline{w}(k))$$

$$+ E(\underline{s}(k))\underline{g}(\underline{x}(k)+\underline{w}(k)) + F\underline{v}(k) \tag{4.25}$$

Hence

$$\|\underline{\Phi}(\underline{x},k)\| \le \max_{\underline{r}\in\mathcal{R}} \|D(\underline{r})\underline{y}\| + \max_{\underline{r}\in\mathcal{R}} \|D(\underline{r})\underline{w}\| + \max_{\underline{w}\in\mathcal{W}} \|G\underline{w}\|$$

$$+ \max_{\underline{s}\in\mathcal{S}} \|E(\underline{s})G\underline{y}\| + \max_{\underline{s}\in\mathcal{S}} \|E(\underline{s})\|\sigma(\underline{y}) + \max_{\underline{v}\in\mathcal{V}} \|F\underline{v}\|$$

$$\triangleq \sigma(\underline{y}) \tag{4.26}$$

In order to justify (4.26), we add up the following assumption:

$$\text{(A8)}\quad 1 - \max_{\underline{s}\in\mathcal{S}} \|E(\underline{s})\| > 0 \tag{4.27}$$

For ease of analysis, we define the following norm quantities:

$$\rho_r = \max_{\underline{r}\in\mathcal{R}} \|D(\underline{r})\| \qquad \rho_s = \max_{\underline{s}\in\mathcal{S}} \|E(\underline{s})\| \tag{4.28a}$$

$$\rho_v = \max_{\underline{v} \in \mathcal{V}} \|F\underline{v}\| \qquad \rho_w = \max_{\underline{w} \in \mathcal{W}} \|\underline{w}(k)\| \tag{4.28b}$$

$$\rho_g = \|G\| \qquad \rho_{sg} = \max_{\underline{s} \in \mathcal{S}} \|E(\underline{s})G\| \tag{4.28c}$$

Using assumption (A8) and the notations in (4.28), one can simply obtain:

$$\sigma(\underline{y}) = [(\rho_r + \rho_{sg})\|\underline{y}\| + (\rho_r + \rho_g)\rho_w + \rho_v]/(1 - \rho_s) \tag{4.29}$$

From (4.21) and (4.29), we define $\sigma(x): R^n \longrightarrow R_+$ by

$$\sigma(\underline{x}) = \alpha + \beta \|\underline{x}\| \tag{4.30}$$

where

$$\alpha = [(2\rho_r + \rho_g + \rho_{sg})\rho_w + \rho_v]/(1 - \rho_s) \tag{4.31a}$$

$$\beta = (\rho_r + \rho_{sg})/(1 - \rho_s) \tag{4.31b}$$

As a consequence of (4.26)-(4.31), it is straight forward to see that $\forall\ \underline{x}, \underline{y}, \ \underline{y} = \underline{x} + \underline{w}$:

$$\sigma(\underline{y}) \le \sigma(\underline{x}) \tag{4.32}$$

In view of the boundedness theorem (4.5.1), we consider a sphere $\mathcal{D}(\mu)$, centered at the origin ($\underline{x}=\underline{0}$) and with radius:

$$\mu = \frac{\beta}{\lambda_m(Q)}\left(\frac{\epsilon}{4} + 2\|B^T P\tilde{A}\|\rho_w\right) + \left[\frac{\beta^2}{\lambda_m(Q)}\left(\frac{\epsilon}{4} + 2\|B^T P\tilde{A}\|\rho_w\right) + \frac{\alpha}{\lambda_m(Q)}\left(\frac{\epsilon}{2} + 4\|B^T P\tilde{A}\|\rho_w\right)\right]^{1/2} \tag{4.33}$$

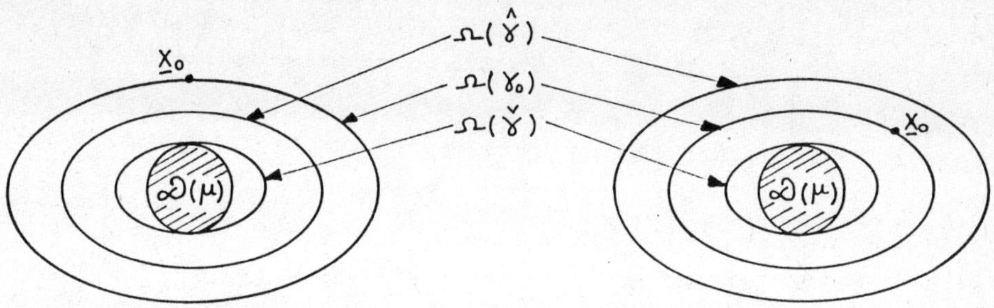

Figure 4.4: Lyapunov Ellipsoids

As in [72], assume that the initial conditions x_o is contained within an ellipsoid $\Omega(\gamma_o)$ with $\gamma_o = \underline{x}_o^T P \underline{x}_o$ and define two ellipsoids $\Omega(\hat{\gamma})$ and $\Omega(\check{\gamma})$ with $\hat{\gamma} > \check{\gamma}$ and $\check{\gamma} = \mu^2 \lambda_m(P)$. These are illustrated in Fig. 4.4. Finally, If $\underline{x} \notin \Omega(\hat{\gamma})$, define a set

$$c_o = \min \{ \underline{x}^T Q \underline{x} - (4\|B^T P\tilde{A}\| \rho_w + \epsilon/2)\sigma(\underline{x}) \; : \; \underline{x} \in \mathcal{D}^c \} \tag{4.34}$$

The role of $\mathcal{D}(\mu)$ will be explained subsequently.

<u>Theorem 4.6.1</u>

Consider system (4.1),(4.21) under the application of control (4.22) and satisfying assumptions (A1-A5,A8). Then, any solution starting from any initial condition $(\underline{x}_o, k_o) \in R^n \times \mathcal{B}$ has the following properties:

(i) bounded in a region of attraction $\mathcal{D}^c$, where $\mathcal{D}^c$ is the complement of $\mathcal{D}$ with radius μ given by (4.33).

(ii) uniformly bounded with

$$\delta(\underline{x}_o) = \begin{cases} \|\underline{x}_o\| \sqrt{\lambda_M(P)/\ \lambda_m(P)} & \text{for } \underline{x}_o \notin \Omega(\hat{\gamma}) \\ \sqrt{\hat{\gamma}/\ \lambda_m(P)} & \text{for } \underline{x}_o \in \Omega(\hat{\gamma}) \end{cases} \tag{4.35}$$

(iii) uniformly ultimately bounded w.r.t. $\Omega(\hat{\gamma})$ with

$$\tilde{K}\ (\underline{x}_o, \Omega(\hat{\gamma})) = \begin{cases} (\gamma_o - \hat{\gamma})/c_o & \text{for } \underline{x}_o \notin \Omega(\hat{\gamma}) \\ 0 & \text{for } \underline{x}_o \in \Omega(\hat{\gamma}) \end{cases} \tag{4.36}$$

<u>Proof</u>

To prove (i), consider a candidate Lyapunov function $V(.) : R^n \longrightarrow R_+$

$$V(\underline{x}(k),k)=\underline{x}^T(k)P\underline{x}(k) \tag{4.37}$$

where P is the solution of (4.4). Given any admissible uncertainties $\underline{r}(.)$, $\underline{s}(.)$, $\underline{v}(.)$ and $\underline{w}(.)$, the Lyapunov difference corresponding to the resulting closed loop system (4.24) and the Lyapunov function (4.37) is given by:

$$\Delta V(\underline{x}(k),k) = V(\underline{x}(k+1)) - V(\underline{x}(k))$$

$$= \underline{x}^T(k+1)P\underline{x}(k+1) - \underline{x}^T(k)P\underline{x}(k) \tag{4.38}$$

A little algebra on (4.4), (4.21), (4.26) and dropping the suffices for simplicity, yields:

$$\Delta V = [\underline{x}^T(k)\tilde{A}^T + \underline{g}^T(\underline{y})B^T + \underline{\Phi}^T(\underline{x},k)B^T\ ]\ P\ [\tilde{A}\ \underline{x}(k) + B\underline{g}(\underline{y}) +$$

$$+ B\underline{\Phi}(\underline{x},k)] - \underline{x}^T(k)P\underline{x}(k)$$

$$\Delta V = \underline{x}^T[\tilde{A}^T P\tilde{A} - P]\underline{x} + 2\underline{x}^T\tilde{A}^T PB[\underline{g}(\underline{y}) + \underline{\Phi}(\underline{x},k)] +$$

$$+ [\underline{g}(\underline{y}) + \underline{\Phi}(\underline{x},k)]^T B^T PB\,[\underline{g}(\underline{y}) + \underline{\Phi}(\underline{x},k)]$$

$$\leq -\underline{x}^T Q\underline{x} + 2(B^T P\tilde{A}\,\underline{y})^T[\underline{g}(\underline{y}) + \frac{B^T P\tilde{A}\,\underline{y}}{\|B^T P\tilde{A}\,\underline{y}\|}\sigma(\underline{y})] -$$

$$- 2(B^T P\tilde{A}\,\underline{w})^T\,[\underline{g}(\underline{y}) - \frac{B^T P\tilde{A}\,\underline{w}}{\|B^T P\tilde{A}\underline{w}\|}\sigma(\underline{y})] +$$

$$+ [\underline{g}(\underline{y}) + \frac{B^T P\tilde{A}\underline{y}}{\|B^T P\tilde{A}\underline{y}\|}\sigma(\underline{y})]^T B^T PB[\underline{g}(\underline{y}) + \frac{B^T P\tilde{A}\underline{y}}{\|B^T P\tilde{A}\underline{y}\|}\sigma(\underline{y})] \quad (4.39)$$

As a consequence of (4.23), the second term on the r.h.s. of (4.39) vanishes for $\|B^T P\tilde{A}\underline{y}\| > \epsilon$, but if $\|B^T P\tilde{A}\underline{y}\| \leq \epsilon$,its maximum value (at $\|B^T P\tilde{A}\underline{y}\| = \epsilon/2$) is equal to $\epsilon\sigma(\underline{y})/2$. The maximum value of the third term occurring for $\|B^T P\tilde{A}\underline{y}\| > \epsilon$ and $B^T P\tilde{A}\underline{w} / \|B^T P\tilde{A}\underline{w}\| = B^T P\tilde{A}\underline{y} / \|B^T P\tilde{A}\underline{y}\|$ is simply $4\,\|B^T P\tilde{A}\underline{w}\|\,\sigma(\underline{y})$. with respect to the fourth term, it vanishes for $\|B^T P\tilde{A}\underline{y}\| > \epsilon$ but if $\|B^T P\tilde{A}\underline{y}\| \leq \epsilon$, its maximum value occurs at $\|B^T P\tilde{A}\underline{y}\| = \epsilon$ lest the term vanishes again. Regrouping the terms and summing up the maxima, we have:

$$\Delta V(k) \leq -x^T Qx + [\epsilon/2 + 4\,\|B^T P\tilde{A}\|\,\mathcal{S}_w]\,\sigma(\underline{x})$$

$$\leq -\{\lambda_m(Q)\|\underline{x}\|^2 - [\epsilon/2 + 4\|B^T P\tilde{A}\|\,\mathcal{S}_w](\alpha + \beta\,\|\underline{x}\|\,\}$$

$$< 0 \quad (4.40)$$

When

$$\lambda_m(Q)\,\|\underline{x}\|^2 - [\epsilon/2 + 4\,\|B^T P\tilde{A}\|\,\mathcal{S}_w]\,(\alpha + \beta\,\|\underline{x}\|) > 0 \quad (4.41a)$$

or equivalently in the light of (4.33)

$$\|\underline{x}\| > \mu \qquad (4.41b)$$

We note that (4.40) or (4.41) is valid for all $(\underline{x},k) \in \mathcal{D}^c \times \mathcal{Z}$.

(ii) In view of (i), let $\underline{x}(.):[k_0,k_1] \longrightarrow R^n$, $\underline{x}(k_0)=\underline{x}_0$ is a solution of closed loop uncertain system (4.24). Hence, we have two cases (refer to Fig. 4.4):

(c1) $\underline{x}_0 \notin \Omega(\hat{\gamma}) \Longrightarrow \underline{x}(k) \in \Omega(\gamma_0) \quad \forall\, k \in [k_0,k_1]$

then

$$0 \le \lambda_m(P)\|\underline{x}(k)\|^2 \le \underline{x}^T(k)P\underline{x}(k) \le \underline{x}_0^T P\underline{x}_0 \le \lambda_M(P)\|\underline{x}_0\|^2 \qquad (4.42a)$$

(c2) $\underline{x}_0 \in \Omega(\hat{\gamma}) \Longrightarrow \underline{x}(k) \in \Omega(\hat{\gamma}) \quad \forall\, k \in [k_0,k_1]$

then

$$0 \le \lambda_m(P)\|\underline{x}(k)\|^2 \le \underline{x}^T(k)P\underline{x}(k) \le \hat{\gamma} \qquad (4.42b)$$

From (4.42a) and (4.42b), it is readily seen that (4.35) is verified and (ii) is proved.

(iii) The result (iii) follows directly from condition (4.41). Here again, we have two cases:

(c1) $\underline{x}_0 \in \Omega(\hat{\gamma})$, then the solution is bounded with

$\tilde{K}(\underline{x}_0,\Omega(\hat{\gamma}))=0$.

(c2) $\underline{x}_0 \notin \Omega(\hat{\gamma})$, then $V(\underline{x}(k))$ decreases as long as $\underline{x}(k) \in \Omega(\hat{\gamma})$ and the solution trajectory $\Delta\Omega(\hat{\gamma})$ in a finite number of periods. An upper bound of this interval is determined by considering the summation:

$$\sum_{j=k_o}^{\hat{k}-1} \Delta V(j) = \Delta V(k_o) + \Delta V(k_o+1) + \ldots\ldots\ldots\ldots + \Delta V(\hat{k}-1)$$

$$= [V(\underline{x}(k_o+1)) - V(\underline{x}(k_o))] + [V(\underline{x}(k_o+2)) - V(\underline{x}(k_o+1))]$$

$$+ \ldots\ldots\ldots\ldots\ldots\ldots + [V(\underline{x}(\hat{k})) - V(\underline{x}(\hat{k}-1))]$$

$$= V(\underline{x}(\hat{k})) - V(\underline{x}(k_o))$$

$$= \hat{\gamma} - \gamma_o \qquad (4.43)$$

However, in view of (4.34) and (4.40), we have:

$$\Delta V(k) \leq - co \qquad (4.44)$$

thus

$$\sum_{j=k_o}^{\hat{k}-1} \Delta V(j) \leq -(\hat{k} - ko)\, co \qquad (4.45)$$

combining (4.43) and (4.45), one obtains

$$\hat{\gamma} - \gamma_o \leq -(\hat{k} - ko) co$$

Therefore,

$$\hat{k} - ko \triangleq \tilde{K} \leq (\gamma_o - \hat{\gamma})/co \qquad (4.46)$$

This concludes the proof of the Theorem 4.6.1. ∎∎∎

<u>Remark 4.6.1</u>

An appropriate selection of the linear part of the controller, namely G, would be the optimal discrete gain to ensure the stabilizability of A [2,70].

Corollary 4.6.1

In the absence of uncertainty ($\underline{w}=\underline{0}$), and if there are infinite feedback gain, that is $\epsilon=0$, then the switching term will be given by [68]:

$$\underline{g}(\underline{y}) = \underline{g}(\underline{x}) \begin{cases} = -\sigma(\underline{x})\,\mathrm{sgn}(B^T P\tilde{A}\underline{x}) & \text{for } \|B^T P\tilde{A}\underline{x}\| \neq 0 \\ \in \{\underline{z}\in R : \|\underline{z}\| \leq \sigma(\underline{x}) & \text{for } \|B^T P\tilde{A}\underline{x}\| = 0 \end{cases} \tag{4.47}$$

It follows from (4.33) in this case that $\mu = 0$ and $\mathcal{D}$ = empty set, that is $\mathcal{D}^c$ will represent the whole space. On the other hand from (4.40),

$$\Delta V \leq -\,\underline{x}^T Q \underline{x} \qquad \forall\ (\underline{x},k)\in R^n \times \mathcal{Z} \tag{4.48}$$

Corollary 4.6.2

If $\underline{y}(k_0)$ is known, then it follows from (4.21) and (4.35) that

$$\delta(\underline{x}_0) = \begin{cases} (\|\underline{y}(k_o)\| + \rho_\omega)\sqrt{\lambda_M(P)/\lambda_m(P)} & \text{for } \underline{x}_o \notin \Omega(\hat{\gamma}) \\ \sqrt{\hat{\gamma}/\lambda_m(P)} & \text{for } \underline{x}_o \in \Omega(\hat{\gamma}) \end{cases} \tag{4.49}$$

4.7 Linear Feedback Control

In order to evaluate the efficacy of nonlinear control (4.22), we consider the reduced linear control law obtained from (4.22) by setting $\underline{g}(\underline{y}) = \underline{0}$, given by :

$$\underline{u}(k) = G_0\ \underline{y}(k) \tag{4.50}$$

where G_0 may be different from G. With the application of the control (4.50) on the uncertain system (4.1),(4.21) the closed loop system is given by:

$$\underline{x}(k+1) = \tilde{A}_0\ \underline{x}(k) + B\underline{\Phi}_0(\underline{x}(k),k) \tag{4.51a}$$

$$\underline{x}(k_0) = \underline{x}_0 \tag{4.51b}$$

where

$$\tilde{A}_0 \triangleq A + B\ G_0 \tag{4.52}$$

$$\underline{\Phi}_0(\underline{x}(k),k) \triangleq D(\underline{r}(k))\underline{x}(k) + G_0\ \underline{w}(k) + E(\underline{s}(k))G_0\ (\underline{x}(k) + \underline{w}(k)) + F\ \underline{v}(k) \tag{4.53}$$

Hence

$$\|\underline{\Phi}_0(\underline{x}(k),k)\| \leq \alpha_0 + \beta_0\ \|\underline{x}\| \triangleq \sigma_0(\underline{x}) \tag{4.54}$$

with

$$\alpha_0 = [\ \|G_0\| + \underset{\underline{s}\in\mathcal{S}}{\text{Max}}\ \|E(\underline{s})G_0\|\ ]\ \rho_w + \rho_v \tag{4.55a}$$

$$\beta_0 = \rho_r + \underset{\underline{s}\in\mathcal{S}}{\text{Max}}\ \|E(\underline{s})G_0\| \tag{4.55b}$$

To investigate the corresponding boundedness behaviour, define a candidate Lyapunov function $V_0(\underline{x}) = \underline{x}^T(k)\ P_0\ \underline{x}(k)$ where P_0 is the solution of Lyapunov equation:

$$\tilde{A}_0^T\ P_0\ \tilde{A}_0 - P_0 = -\ Q_0 \quad ; \quad Q_0 > 0 \tag{4.56}$$

then

$$\Delta V_o(k) = \underline{x}^T(k+1)\, P_o\, \underline{x}(k+1) - \underline{x}^T(k)\, P_o\, \underline{x}(k)$$

$$= \underline{x}^T(k)[\, \tilde{A}_o^T\, P_o\, \tilde{A}_o - P_o\,]\underline{x}(k) + 2\, \underline{x}^T\, \tilde{A}_o^T\, P_o\, B\, \underline{\Phi}_o(\underline{x},k)$$

$$+ \underline{\Phi}_o^T(\underline{x},k)\, B^T\, P_o\, B\, \underline{\Phi}_o(\underline{x},k)$$

$$\le -\, \underline{x}^T\, Q_o\, \underline{x} + 2\, \|B^T\, P_o\, \tilde{A}_o\, \underline{x}\|\, \sigma_o(\underline{x}) + \lambda_M(B^T\, P_o\, B)\, \sigma_o(\underline{x}) \tag{4.57}$$

We observe that $\Delta V_o(k) < 0$ if the following condition is satisfied:

$$\underline{x}^T\, Q_o\, \underline{x} - 2\|B^T\, P_o\, \tilde{A}_o\|\cdot\|\underline{x}\|\, (\alpha_o + \beta_o\, \|\underline{x}\|) -$$

$$-\lambda_M(B^T\, P_o\, B)\, [\alpha_o + \beta_o\, \|\underline{x}\|]^2 > 0 \tag{4.58}$$

Alternatively,

$$[\lambda_m(Q_o) - 2\beta_o\|B^T\, P_o\, \tilde{A}_o\| - \beta_o^2\, \lambda_M(B^T\, P_o\, B)]\|\underline{x}\|^2 - 2[\alpha_o\|B^T\, P_o\, \tilde{A}_o\|$$

$$+ \alpha_o\, \beta_o\, \lambda_M(B^T\, P_o\, B)]\|\underline{x}\| - \alpha_o^2\, \lambda_M(B^T\, P_o\, B) > 0 \tag{4.59}$$

which can only hold when

$$\lambda_m(Q_o) - 2\beta_o\, \|B^T\, P_o\, \tilde{A}_o\| - \beta_o^2\, \lambda_M(B^T\, P_o\, B) > 0 \tag{4.60}$$

As a result, provided that condition (4.60) is satisfied, it occurs that $\Delta V_o(k) < 0$ for all $(\underline{x},k) \in \mathcal{D}_o^c \times \mathcal{Z}$ where $\mathcal{D}_o^c$ is the complement of the sphere $\mathcal{D}_o$ with radius given by:

$$\mu_o = [\alpha_o\, \|B^T\, P_o\, \tilde{A}_o\| + \alpha_o\, \beta_o\, \lambda_M(B^T\, P_o\, B)]/\phi + \{[\alpha_o\, \|B^T\, P_o\, \tilde{A}_o\|$$

$$+ \alpha_o\, \beta_o\, \lambda_M(B^T\, P_o\, B)]^2 + \alpha_o^2\, \lambda_M(B^T\, P_o\, B)\}^{1/2} / \phi \tag{4.61a}$$

where

$$\phi = \lambda_m(Q_0) - 2\beta_0 \|B^T P_0 \tilde{A}_0\| - \beta_0^2 \lambda_M(B^T P_0 B) \tag{4.61b}$$

Finally, as above, define a set

$$\bar{c}_0 = \min \{\underline{x}^T Q_0 \underline{x} - 2\|B^T P_0 \tilde{A}_0 \underline{x}\|\sigma_0(\underline{x}) - \lambda_M(B^T P_0 B)\sigma_0^2(\underline{x}) : \underline{x} \in \mathcal{D}_o^c \} \tag{4.62}$$

Theorem 4.7.1

Consider the uncertain system (4.1),(4.21) with the application of control (4.50) and taking into consideration the assumption (A1-A5,A8), and condition (4.60). Then, any solution starting from any initial condition $(\underline{x}_0, k_0) \in R^n \times \mathcal{J}$ is

(i) bounded with respect to region $\mathcal{D}_o^c$, which is the complement of $\mathcal{D}_o$ with radius μ_0 given by (4.61).

(ii) Uniformly bounded with

$$\delta(\underline{x}_0) = \begin{cases} \|\underline{x}_0\| \sqrt{\lambda_M(P_0)/\lambda_m(P_0)} & \text{for } \underline{x}_0 \notin \Omega(\hat{\gamma}) \\ \sqrt{\hat{\gamma}/\lambda_m(P_0)} & \text{for } \underline{x}_0 \in \Omega(\hat{\gamma}) \end{cases} \tag{4.63}$$

(iii) Uniformly ultimately bounded w.r.t $\Omega(\hat{\gamma})$ with

$$\tilde{K}(\underline{x}_0, \Omega(\hat{\gamma})) = \begin{cases} (\gamma_0 - \hat{\gamma})/c_0 & \text{for } \underline{x}_0 \notin \Omega(\hat{\gamma}) \\ 0 & \text{for } \underline{x}_0 \in \Omega(\hat{\gamma}) \end{cases} \tag{4.64}$$

Proof

Follow parallel development to Theorem 4.6.1.

■■■

Remark 4.7.1

For the purpose of comparison , let $G = G_o$ and $Q = Q_o$, which imply that $P = P_o$. In terms of the radii μ and μ_o, given by (4.33) and (4.61 respectively , one can simply observe that $\mu < \mu_o$ and hence $\mathcal{D}^c > \mathcal{D}_o^c$. This interesting observation adds superiority for using nonlinear controller rather than linear one.

Corollary 4.7.1

In view of the above remark and in addition , assume the special case $\epsilon=0$ and $\rho_w=0$. This results in $\mu \longrightarrow 0$ and $\mu_o \longrightarrow$ constant value > 0 , that is the uniform asymptotic stability behaviour can be guaranteed via nonlinear control again rather than the linear one.

Next, we demonstrate the theoretical results by a typical system.

4.8 Illustrative Example

We now present the computer simulation results of a control system. The standard open-loop version of the Ward-Leonard speed controller [73] is described by:

$$\dot{\underline{x}}(t) = \begin{bmatrix} -R_f/L_f & 0.0 & 0.0 \\ k_g/L_a & -R_a/L_a & -k_m/L_a \\ 0.0 & k_m/J & -F/J \end{bmatrix} \underline{x}(t) + \begin{bmatrix} 1/L_f & 0.0 \\ 0.0 & 0.0 \\ 0.0 & 1/J \end{bmatrix} \underline{u}(t)$$

where the state vector $\underline{x}$ consists of the field current x_1 , armature current x_2 and the load angle velocity x_3. Two inputs are used: u_1 is the voltage applied across the generator field winding and u_2 is the torque acting on the load. The choice of parameter values $R_f = 10\ \Omega$, $L_f = 80$ H, $K_g = 100$ V/A, $R_a = 1.8\ \Omega$, $L_a = 10$ H, $k_m = 3$ V rad sec., $J = 6$ Kg m^2, $F = 7.5$ Nm rad$^{-1}\cdot$sec^{-1}, together with a discretization scheme with time increment of $\Delta t = 0.05$ sec., yields the matrices of the system (4.1) of the form [5]:

$$A = \begin{bmatrix} 0.797 & 0.0 & 0.0 \\ 0.5127 & 0.793 & -0.0154 \\ 0.0 & 0.0145 & 0.764 \end{bmatrix}, \quad B = \begin{bmatrix} 0.00699 & 0.0 \\ 0.0001271 & -0.0000507 \\ 0.0 & 0.0932 \end{bmatrix}$$

We consider that the parameters R_a, L_f and J undergo ±10% variation about their nominal values. This means that the elements a_{11}, a_{22}, a_{32}, a_{33}, b_{11}, b_{32} of the system matrices are allowed to change by ± 10% whereas the other elements are kept at their nominal values. We thus have

$$\Delta A(\underline{r}) = \begin{bmatrix} r_1 & 0.0 & 0.0 \\ 0.0 & r_2 & 0.0 \\ 0.0 & r_3 & r_4 \end{bmatrix}, \quad \Delta B(\underline{s}) = \begin{bmatrix} s_1 & 0.0 \\ 0.0 & 0.0 \\ 0.0 & s_2 \end{bmatrix}$$

and the compact bounding sets $\mathcal{R}$ and $\mathcal{S}$ are given by

$$\mathcal{R} = \{\ \underline{r} \in R^4\ ;\ |r_1| \le 0.0797\ ,\ |r_2| \le 0.0793\ ,\ |r_3| \le 0.00145\ ,$$
$$|r_4| \le 0.0764\ \}$$

$$\mathcal{S} = \{ \underline{s} \in R^2 \; ; \; |s_1| \leq 0.000699 \; , \; |s_2| \leq 0.00932 \}$$

It is worth mentioning that the matching conditions are met with

$$D(\underline{r}) = \begin{bmatrix} 143.0143\, r_1 & 2.6004\, r_2 + 1.4146\text{x}10^{-3}\, r_3 & 1.4146\text{x}10^{-3}\, r_4 \\ 1.0609\text{x}10^{-4}\, r_1 & -5.8348\text{x}10^{-3}\, r_2 + 10.7296\, r_3 & 10.72609\, r_4 \end{bmatrix}$$

$$E(\underline{s}) = \begin{bmatrix} 143.01434\, s_1 & 1.41459\text{x}10^{-3}\, s_2 \\ 1.06094\text{x}10^{-4}\, s_1 & 10.729609\, s_2 \end{bmatrix}$$

In our simulation, we took

$$G = \begin{bmatrix} -0.0741 & -0.0167 & 0.0012 \\ 0.0127 & -0.0017 & -0.1633 \end{bmatrix}$$

Thus

$$\tilde{A} \;\hat{=}\; A + BG = \begin{bmatrix} 0.7965 & -0.0001 & 0.0 \\ 0.5127 & 0.793 & -0.0154 \\ 0.0012 & 0.0143 & 0.7488 \end{bmatrix}$$

With $Q = I_3$, the solution of (4.4) is given by :

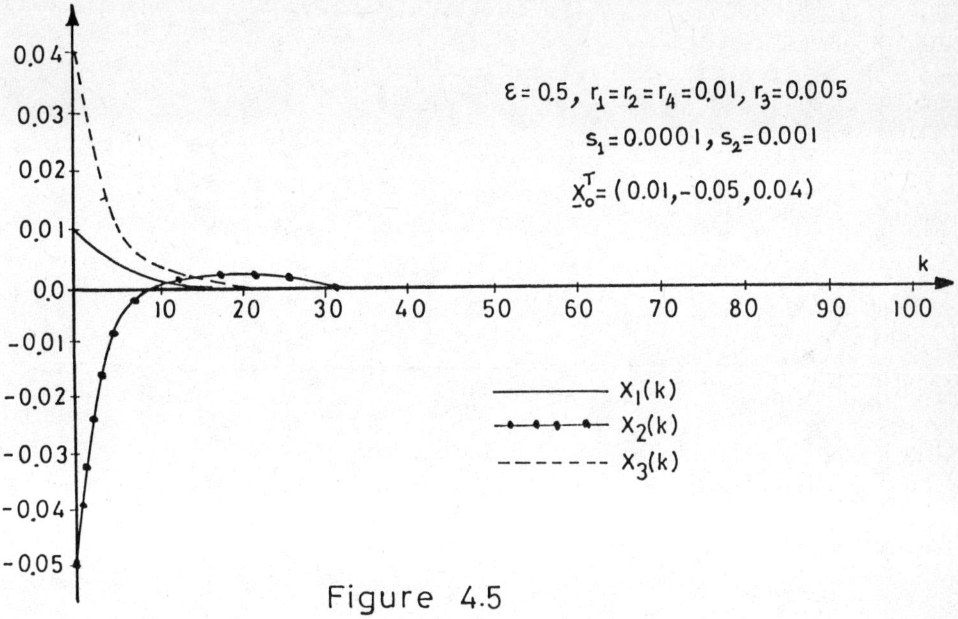

Figure 4.5

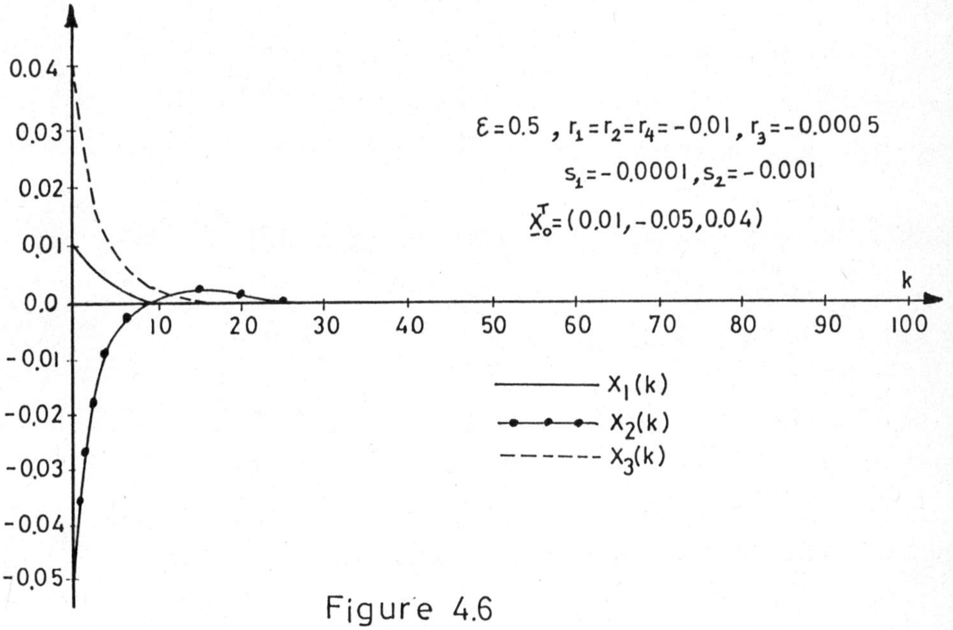

Figure 4.6

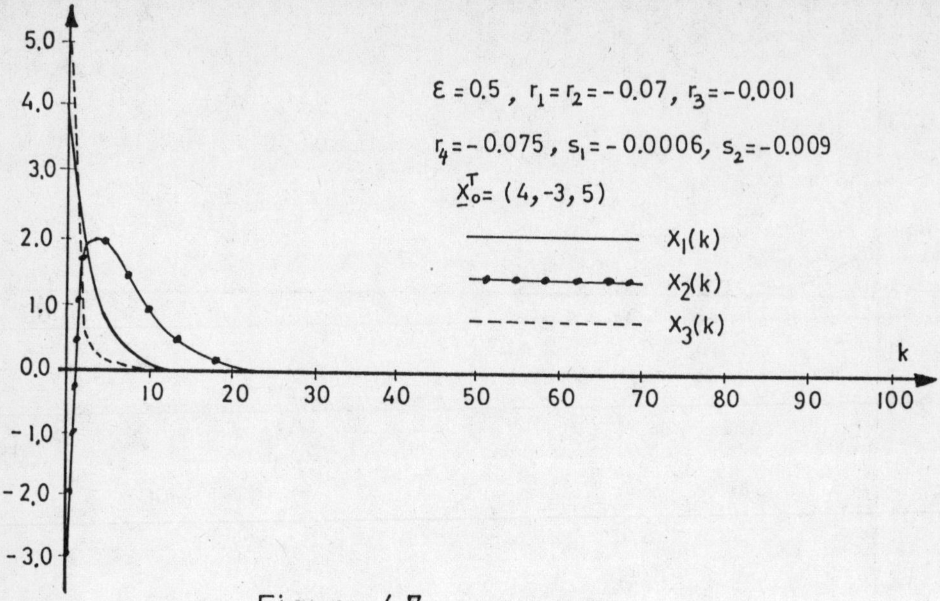

Figure 4.7

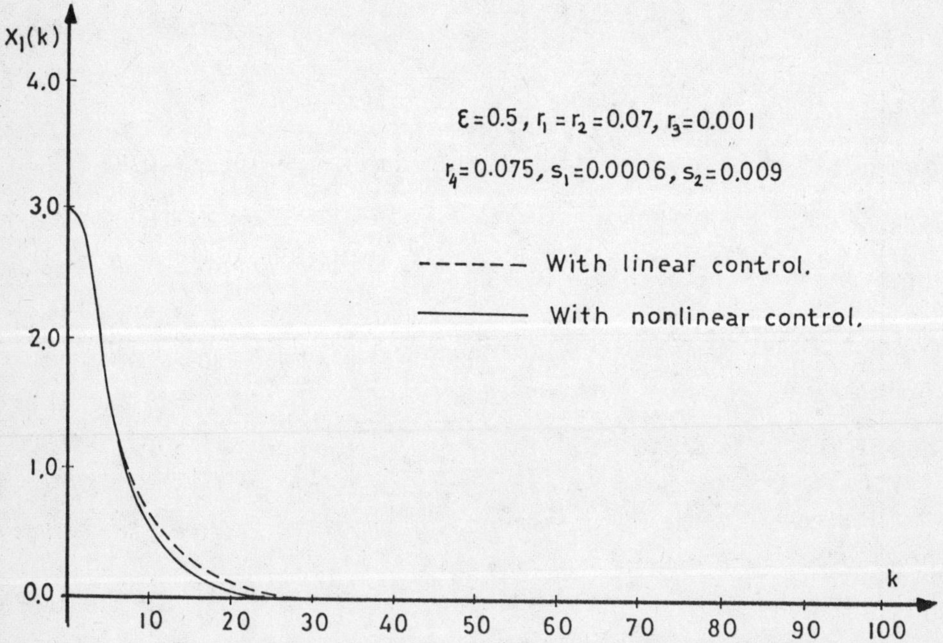

Figure 4.8

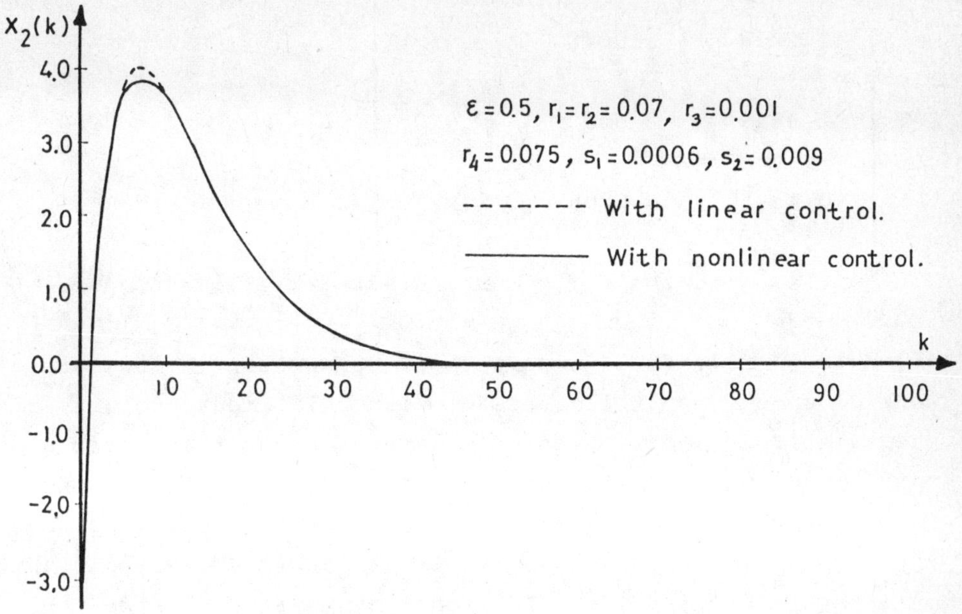

Figure 4.9

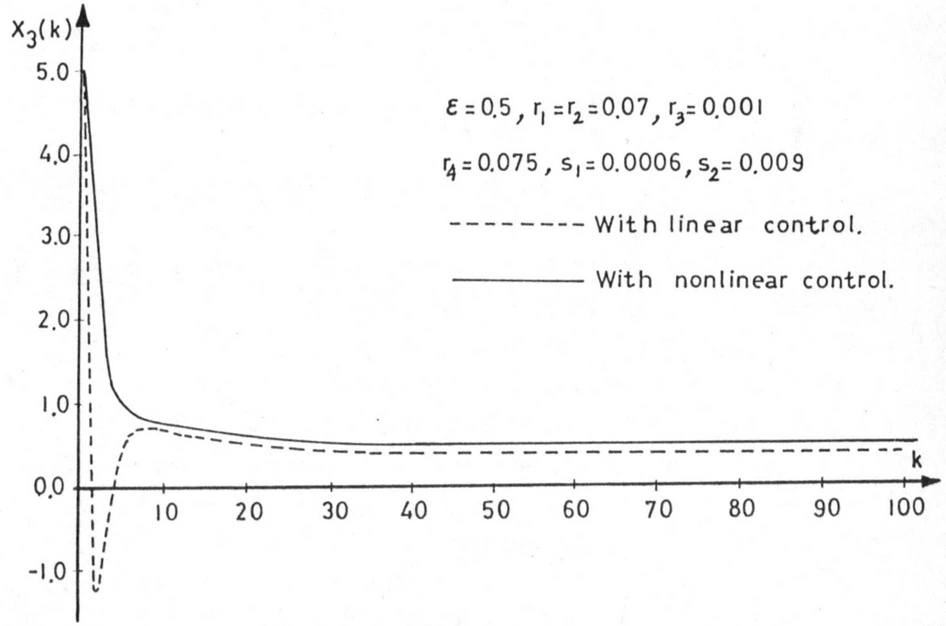

Figure 4.10

$$P = \begin{bmatrix} 11.2928 & 2.9642 & -0.1575 \\ 2.9642 & 2.693 & -0.0208 \\ -0.1575 & -0.0208 & 2.2789 \end{bmatrix}$$

It is noted that $\rho_s = \underset{\underline{s} \in S}{\mathrm{Max}} \; \|E(\underline{s})\| = 0.1$, i.e assumption (A8) holds. Now, for the purpose of comparison , the Ward - Leonard speed controller system with different values of uncertainties, has been simulated using the proposed controllers and the resultant closed loop state trajectories are indicated in Figs. 4.5-4.10. From these simulations, one can see that the closed loop system is stable. Moreover , the nonlinear control policy stabilizes the system faster than the purely linear control and this adds superiority for using the nonlinear control scheme rather than linear one.

4.9 Conclusions

The main result of this Chapter shows that the stabilization of linear time - invariant , discrete systems with additive - type bounded uncertainty can be achieved by a nonlinear as well as linear feedback control schemes. The two strategies resulted in stable closed loop system. However , from the derived analysis and simulation results , one must have to distinguish between two options :

(a) use of nonlinear control structure which ensures uniform asymptotic stability or uniform ultimate boundedness behaviour for a wide class of bounded admissible uncertainties . But it is complex to be realized.

(b) or use an easy implemented linear feedback control policy . This , however gives domain of stability relatively smaller in comparison with (a) above.

CHAPTER 5

MULTIPLE-CONTROLLER SCHEMES FOR DISCRETE SYSTEMS

5.1 Introduction

The main task of control engineering design is to ensure that the dynamic system under consideration behaves in some desirable way. In Chapter 4, we considered the feedback design problem of discrete-time uncertain dynamical systems, and developed new approaches for synthesizing controllers which lead to either asymptotic stability or ultimate boundedness of the state of such systems. However, the proposed controllers suffer from the following :

(1) As it has been shown from the derived analysis, the so-called "matching conditions" are assumed to be satisfied. It is well known that these matching assumptions constitute sufficient conditions for a given uncertain system to be stabilized. However, even for uncertain linear systems the matching conditions are known to be undully restrictive. Indeed, it has been shown in [42,44] that there exist many uncertain linear systems which fails to satisfy the matching conditions and yet are nevertheless stabilizable.

(2) The developed controllers require access state vector to the system for implementation. However, in many cases, it may be impossible to measure all of the states of the system .

Realizing these drawbacks, much effort have directed towards developing control methodologies in order to stabilize a larger class of systems than those which satisfy the matching conditions. Furthermore, it is necessary to design controllers for uncertain systems with incomplete state measurements

The material covered in this Chapter is organized into six basic sections. In section 5.2, we try to rlax the matching conditions which are assumed to be satisfied in the previous Chapter. Instead, the stabilization problem is tackled via developing two-level control scheme [75]. This method is applied to an illustrative example in section 5.3. The simulation results have shown the applicability and feasibility for practical purposes. In section 5.4, we consider an observer-based nonlinear control scheme. The standard full-order observer and the closed-loop state equations are derived in section 5.5. It has been shown, in section 5.6, that the proposed structure is feasible provided that the uncertain parameters don't exceed a certain computable threshold. Finally, the scheme is applied to a typical example to show the effectivness of the control design. Further properties are investigated to explore the potential of these control stabilization methods.

5.2 Two-Level Control Structure

The objective of this section is to develop a two-level control scheme which can be used to solve the problem at hand. The proposed controller is so designed to maintain the optimal behaviour of the system under normal operating conditions and at the same time guarantee both reliability and stability of the system under a wide class of model uncertainty. In order to fulfill these requirements, a feedback controller structure is developed and decomposed of two main parts: an optimal part to move the eigenvalues of the nominal system into the unit disk and a corrective part to ensure the uniform asymptotic stability of the system in presence of parameter uncertainties. We take note that the proposed design method don't require the validity of the matching conditions.

5.2.1 Formulation of The Problem

Consider a class of uncertain dynamical systems shown in Figure 5.1 and described by the following state and output equations:

$$\underline{x}(k+1) = [A + \Delta A(\underline{r}(k))]\ \underline{x}(k) + B\ \underline{u}(k) \tag{5.1}$$

$$\underline{y}(k) = C\ \underline{x}(k) \tag{5.2}$$

where $\underline{x}(k) \in R^n$ is the state; $\underline{u}(k) \in R^m$ is the control, $m<n$; $\underline{y}(k) \in R^l$ is the output and $\underline{r}(k) \in R^p$ is the vector of uncertain parameters. A, B and C are constant matrices of appropriate dimensions. $\Delta A(r)$ is the model uncertainty matrix and $k \in \mathcal{Z}$ where $\mathcal{Z}$ is a set of all integer numbers.

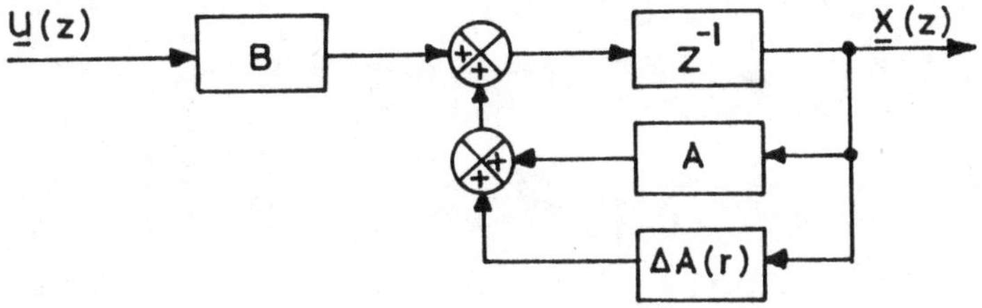

Figure 5.1: Uncertain Discrete System.

In the development to follow, we suppose that the following assumptions hold:

(i) The entries of $\Delta A(.)$ depend (in a continuous manner) on $\underline{r}(k)$ and $\Delta A(\tilde{\underline{r}}) = 0$ for some $\tilde{\underline{r}} \in \mathcal{R}$.

(ii) The function $\underline{r}(.)$ is restricted to be Lebesgue measurable function in a compact bounding set

$\mathcal{R} = \{\underline{r} \in R^p : |r_i| \leq \hat{r}$ for $i=1,2,..,p\}$, $\hat{r}$ is a given constant.

(iii) The triplet {A,B,C} is both completely reachable and reconstructable.

The objective is to develop a multi-controller structure that guarantees the asymptotic stability for a class of uncertain systems in the framework described above.

5.2.2 <u>The Nominal System</u>

We start by considering the nominal system in which the uncertain term $\Delta A(\underline{r}(k))$ is ident ically zero. We assume that with this system is an associated quadratic cost

$$J = 1/2 \sum_{k=0}^{\infty} \underline{y}^T(k)\, Q\, \underline{y}(k) + \underline{u}^T(k)\, R\, \underline{u}(k) \tag{5.3}$$

where $Q = Q^T \geq 0$ and $R = R^T \geq 0$ are weighting matrices. The nominal optimal control $\underline{u}^*(k)$ that minimizes J subject to the constraints

$$\underline{x}(k+1) = A\, \underline{x}(k) + B\, \underline{u}(k) \tag{5.4}$$

$$\underline{y}(k) = C\, \underline{x}(k) \tag{5.5}$$

is given by [2,76] :

$$\underline{u}^*(k) = -\, R^{-1}\, B^T\, P\, (I + B\, R^{-1}\, B^T\, P)^{-1}\, A\, \underline{x}(k)$$

$$= -\, G^*\, \underline{x}(k) \tag{5.6}$$

where P is the positive semidefinite solution of the discrete algebraic Riccati equation

$$P = C^T Q C + A^T P (I + B R^{-1} B^T P)^{-1} A \qquad (5.7)$$

where I is the (nxn) identity matrix.

It should be emphasized that the optimal control (5.6) provides a basic regulation task of stabilizing the uncertain system (5.1) and (5.2).

5.2.3 Multi-Controller Structure

We now consider the uncertain system (5.1) and (5.2) with the satisfaction of assumptions (i)-(iii). In view of the fact that the difference between (5.1) and (5.4) is the uncertain pattern, we define

$$\underline{u}(k) = \underline{u}^*(k) + \underline{u}_{c1}(k) + \underline{u}_{c2}(k) \qquad (5.8)$$

where $\underline{u}^*(k)$ is given by (5.6), $\underline{u}_{c1}(k)$ and $\underline{u}_{c2}(k)$ are corrective control components to handle the effect of uncertainties.

Let us fix the uncertain term at known sequence, that is:

$$\Delta A(\underline{\tilde{r}}(k)) = A_p \quad ; \quad \underline{\tilde{r}}(k) \in \mathcal{R} \qquad (5.9)$$

The substitution of (5.8),(5.9) into (5.1), using (5.6) leads us to :

$$\underline{x}(k+1) = (A - B\, G^*)\underline{x}(k) + A_p \underline{x}(k) + B\underline{u}_{c1}(k) + B\underline{u}_{c2}(k) \qquad (5.10)$$

To determine $\underline{u}_{c1}(k)$, we consider the effect of the term $A_p \cdot \underline{x}(k)$ on the system (5.10) as harmful causing undesirable deviations. In order to reduce these undesirable deviations, we choose

$$\underline{u}_{c1}(k) = - M_1 \ \underline{x}(k) \tag{5.11}$$

where $M \in R^{m \times n}$ is the first component of the required corrective signals. It is important to note that the matrix $(A_p - B M_1)$ depends on A_p only. This matrix reduces to the null matrix if and only if [84] :

$$\text{Rank } (B) = \text{Rank } (B \ A_p) = m \tag{5.12}$$

in which case the gain matrix M_1 is given by :

$$M_1 = (B^T \ B)^{-1} \ B^T \ A_p \tag{5.13}$$

In practice, the rank condition (5.12) is rarely satisfied and there is a difference between A_p and $\Delta A(\underline{r})$ in general. Thus, a residual uncertainty term will arise and take the form :

$$H = [\Delta A(r) - A_p] + (A_p - B \ M_1) \tag{5.14}$$

we now turn to the term $B \ \underline{u}_{c2}(k)$. In order to provide for some improvement in the overall feedback scheme, we select :

$$\underline{u}_{c2}(k) = - M_2 \ \underline{x}(k) \tag{5.15}$$

$$\text{and} \quad B\underline{u}_{c2}(k) = - \tilde{B} \ G^* \ \underline{x}(k) \tag{5.16}$$

In view of (5.12)-(5.16), (5.1) can be written as

$$\underline{x}(k+1) = (A + H) \ \underline{x}(k) - (B + \tilde{B}) \ G^* \ \underline{x}(k) \tag{5.17}$$

where $\tilde{B} \in R^{n \times m}$ is the required perturbation in B.

Now, our problem can be identified as the determination of $\tilde{B}$ and hence M_2, $\underline{u}_{c2}$ which would guarantee the asymptotic stability behaviour of the uncertain closed loop system (5.17) without causing any change in the feedback gain matrix G^*. An answer to this problem is given in the following Theorem.

Theorem 5.2.1

Assuming A is nonsingular, let $S \in R^{n \times n}$ is the symmetric, positive semidefinite solution of

$$S = C^T QC + (A+H)^T S \, [I + (B+\tilde{B})R^{-1}(B+\tilde{B})^T S]^{-1} \, (A+H) \tag{5.18}$$

then a perturbation B is given by

$$\tilde{B} = (A+H) \, [S - C^T QC]^{-1} \, (P - C^T QC) \, A^{-1} \, B - B \tag{5.19}$$

results in a stabilizing control

$$\underline{u}(k) = - (G^* + M_1 + M_2) \, \underline{x}(k) \tag{5.20}$$

where M_1 and M_2 are given by (5.13) and (5.16) respectively.

Proof

Replacing $-G^* \, \underline{x}(k)$ in (5.17) by $\underline{\tilde{u}}(k)$, we have

$$\underline{x}(k+1) = (A - B\,G^*) \, \underline{x}(k) + (B + \tilde{B}) \, \underline{\tilde{u}}(k) \tag{5.21}$$

In the light of the analysis of section 5.2.2, the optimal control $\underline{\tilde{u}}^*(k)$ that minimizes J in (5.3) subject to the constraints (5.21) and (5.2) is given by

$$\underline{\tilde{u}}^*(k) = -R^{-1}(B+\tilde{B})^T \, S \, [I+(B+\tilde{B})R^{-1}(B+\tilde{B})^T \, S]^{-1}(A+H) \, \underline{x}(k) \tag{5.22}$$

where S is the positive semidefinite solution of (5.18). From (5.6) and (5.7), the feedback gain can be expressed as

$$G^* = R^{-1} B^T A^{-T} (P - C^T Q C) \qquad (5.23)$$

and from (5.22), (5.15) the corresponding gain is given by :

$$\tilde{G}^* = R^{-1} (B + \tilde{B})^T (A + H)^{-T} (S - C^T Q C) \qquad (5.24)$$

By setting $\tilde{G}^* = G^*$, we get (5.19). In view of (5.13), (5.16), (5.17) and comparing with (5.1), one can simply get (5.20) which completes the proof. ■■■

Had we followed another route based on [2], we could arrive at an alternative simple form of B using the equality

$$R^{-1} (B + \tilde{B})^T S = R^{-1} B^T P \qquad (5.25)$$

which implies that

$$\tilde{B} = S^{-1} P B - B \qquad (5.26)$$

Now, Define

$$D = P - S \; ; \quad E = P - C^T Q C \; ; \quad F = S - C^T Q C \qquad (5.27)$$

Manipulating (5.7), (5.18) and (5.25)-(5.27) together, we arrive at

$$F (A + H)^{-1} - E A^{-1} = (A+H)^T (D+P) - A^T P \qquad (5.28)$$

which can be simplified with the aid of (5.27) to get explicit expression of D :

$$D = (A+H)^T D (A+H) - H P (A+H) + (C^T Q C - P) A^{-1} H \qquad (5.26)$$

Note that the above expression is independent of S and this avoids the solution of (5.18).

To summarize, the design procedure is given by the following algorithm :

Step 1: Solve (5.7) for the Riccati matrix P and use it in (5.6) to compute the feedback gain G^*.

Step 2: Use (5.13) to compute the gain M_1 and from (5.1), (5.14) obtain the uncertain matrix H.

Step 3: From (5.18) and (5.25) we obtain

$$S = C^T QC + (A+H)^T S [I + S^{-1} PB R^{-1} B^T P]^{-1} (A+H) \qquad (5.30)$$

which can be solved to yield the S matrix.

Step 4: The B matrix is computed from (5.26) which is used in (5.16) to obtain the gain matrix M_2.

Step 5: Implement the controller structure (5.20) to the system (5.1) and (5.2) in order to form the closed-loop system and stop.

5.3 Example and Conclusions

A simple second-order model of a stirred-tank [76] is written in the format (5.1) and (5.2) with

$$A = \begin{bmatrix} .9512 & 0 \\ 0 & .9048 \end{bmatrix}, \quad B = \begin{bmatrix} 4.877 & 4.877 \\ -1.1895 & 3.569 \end{bmatrix}$$

$$C = \begin{bmatrix} 0.01 & 0 \\ 0 & 1.0 \end{bmatrix}, \quad \Delta A(\underline{r}) = \begin{bmatrix} r_1 & 0 \\ 0 & r_2 \end{bmatrix}$$

and $\mathcal{R} = \{ \underline{r} \in R^2 : |r_1| \le .19024, \ |r_2| \le .18096 \}$

With $Q = \begin{bmatrix} 50.0 & 0 \\ 0 & 0.02 \end{bmatrix}, \quad R = \begin{bmatrix} 1/3 & 0 \\ 0 & 3 \end{bmatrix},$

Let $\Delta A(\underline{r}) = \begin{bmatrix} 0.17 & 0 \\ 0 & 0.12 \end{bmatrix}, \quad A_p = \begin{bmatrix} 0.19 & 0 \\ 0 & 0.18 \end{bmatrix}$

and running the above algorithm, we have

$$G^* = \begin{bmatrix} 0.07125 & -0.07029 \\ 0.01357 & 0.04548 \end{bmatrix}, \quad M_1 = \begin{bmatrix} 0.02922 & -0.03783 \\ 0.00974 & 0.03783 \end{bmatrix}$$

and $M_2 = \begin{bmatrix} 0.00011 & -.00285 \\ 0.0066 & .01739 \end{bmatrix}$

For the purpose of comparison, the uncertain system was simulated from the initial condition $\underline{x}(0) = (.2 \quad .1)^T$ and the closed loop state and control trajectories are shown in Figures 5.2 and 5.3 respectively. From these results, it can be concluded that the designed controller maintains the asymptotic stability of the uncertain system at hand. It is interesting to note that this controller structure doesn't require the validity of matching conditions.

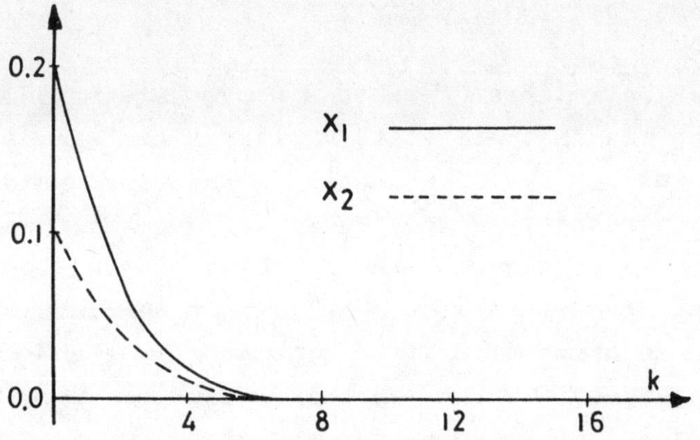

Fig. 5.2: Closed-loop state trajectories of Example 5.3.

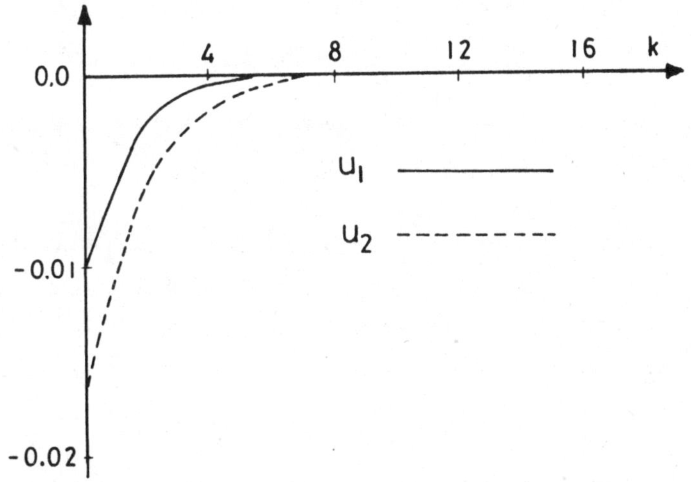

Fig. 5.3: The control trajectories of Example 5.3.

5.4 Observer-Based Nonlinear Control

All the controllers developed above are predicated under the assumption that all state variables are accessible for direct measurement. Unfortunately, this assumption is unrealistic in practice. A well known approach to overcome this difficulty is to generate the feedback control law via an estimate of the state vector [77]. The estimation is performed using an asymptotic state estimator, often called Luenberger observer, which employs only the available directly measurable input and output signals.

At the present time, many techniques for full-order state reconstructor design of linear continuous-time systems exist [76-78]. For discrete-time systems the design procedures of state reconstruction has been dealt with in a similar way as for continuous time systems. However, as pointed out in [2,79,80] ,the one-to-one correspondence between continuous and discrete systems from observer design standpoint is not valid. The main reason for this is that whereas the measurement in continuous-time control systems is instantaneous, there may be a delay (at least one step) in case of discrete-time configurations. Sources of delay are either due to discretization schemes or due to limited information processing capabilities.

In the context of continuous-time, uncertain dynamical systems, some recent works have already dealt with the problem of observing the state variables from the available input-output measurements [81,48,82]. In [81], a measure of robustness is given for an uncertain system stabilized via an observer and a linear controller using only the nominal matrices in constructing the controller/observer pair. The approach taken in [48] requires one to solve a pair of matrix Riccati equations. If their solutions are positive-definite and satisfy a matrix inequality , then there exist observer and feedback control laws which together stabilize the uncertain system. It has been shown in [82] that if some inequalities are satisfied, then one can construct a controller and observer which guarantee a nonzero "measure of robustness".

The objective here is to develop a robust stabilization scheme for a general class of uncertain discrete systems containing uncertain parameters in the system model, in the input connection, in the output distribution and in the direct transmission path of the system. The uncertainties of the system are assumed to be contained within given compact bounding sets. The design approach consists of two steps : First build a state reconstructor depending on the basis of the dynamics of the nominal system; that is the model, obtained by fixing the uncertain parameters at some nominal values, is used as input to the design. Second, the observer takes the control law as given; this control law is designed under the hypothesis of full state feedback [68-70]. The physical realization of the control law,however processes the estimated state instead of the true state. It is worth mentioning that the proposed observer-based nonlinear controller is feasible provided that the uncertain parameters of the system don't exceed a certain computable threshold.

Consider a class of uncertain dynamical systems shown in Fig. 5.4 and described by the following difference equations :

$$\underline{x}(k+1) = [A + \Delta A(\underline{r}(k))]\ \underline{x}(k) + [B + \Delta B(\underline{s}(k))]\ \underline{u}(k) \qquad (5.31)$$

$$\underline{y}(k) = [C + \Delta C(\underline{v}(k))]\ \underline{x}(k) + [D + \Delta D(\underline{w}(k))]\ \underline{u}(k) \qquad (5.32)$$

where $\underline{x}(k) \in R^n$ is the state; $\underline{u}(k) \in R^m$ is the control; $\underline{y}(k) \in R^p$ is the measured output, $p < n$; $\underline{r}(k) \in \mathcal{R} \subset R^a$ is the model uncertainty ; $\underline{s}(k) \in \mathcal{S} \subset R^b$ is the input connection uncertainty; $\underline{v}(k) \in \mathcal{V} \subset R^c$ is the output distribution uncertainty and $\underline{w}(k) \in \mathcal{W} \subset R^d$ is the direct transmission uncertainty. Moreover, A,B,C and D are constant matrices and $\Delta A(.)$, $\Delta B(.)$, $\Delta C(.)$ and $\Delta D(.)$ are matrix functions of appropriate dimensions. $k \in \mathcal{Z}$, where $\mathcal{Z}$ is a set of all integer numbers.

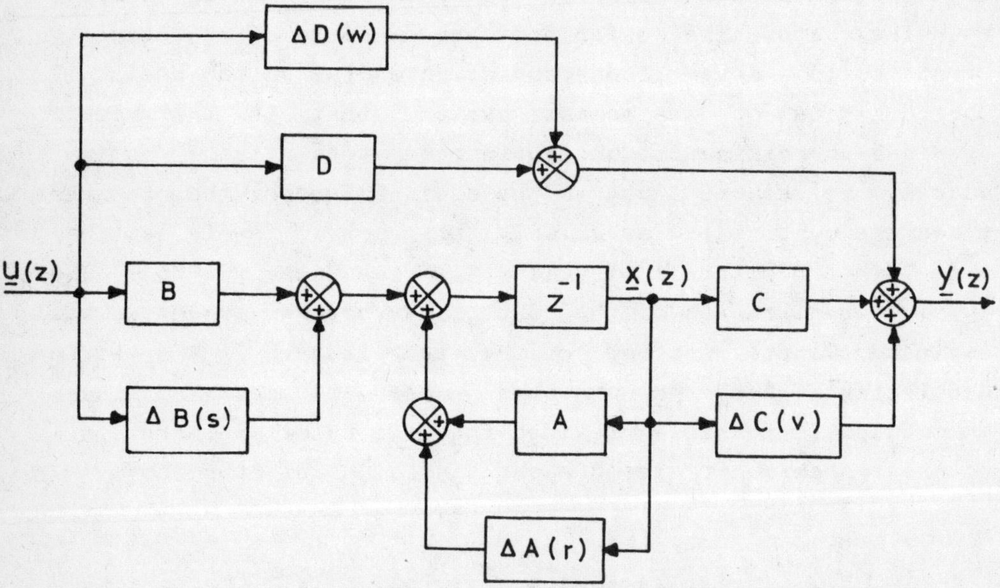

Fig. 5.4: A Generalized uncertain discrete system.

In the sequel, the following assumptions will be taken as standard :

(A1) The entries of $\Delta A(.)$, $\Delta B(.)$, $\Delta C(.)$ and $\Delta D(.)$ are continuous on R^a, R^b, R^c and R^d respectively.

(A2) Uncertainty parameters $\underline{r}(.):R \longrightarrow \mathcal{R}$, $\underline{s}(.):R \longrightarrow \mathcal{S}$, $\underline{v}(.):R \longrightarrow \mathcal{V}$, and $\underline{w}(.):R \longrightarrow \mathcal{W}$ are Lebesgue measurable in compact bounding sets $\mathcal{R}$, $\mathcal{S}$, $\mathcal{V}$ and $\mathcal{W}$ respectively.

(A3) {A,C} is reconstructable pair.

(A4) There exists a class of cone-bounded feedback control function $\underline{g}(.):R^n \longrightarrow R^m$ and an nxn positive-definite symmetric matrix P_c such that the following condition is satisfied : Given any admissible uncertainties $\underline{r}(.)$ and $\underline{s}(.)$, the maximal Lyapunov forward difference corresponding to the closed loop system (5.31) [with control $\underline{u}(k) = \underline{g}(\underline{x}(k))$] and the Lyapunov function $V_c(x) = \underline{x}^T P_c \underline{x}$ satisfies the inequality:

$$\Delta V_{max}(\underline{x}) = \underline{x}^T \{[A + \Delta A(\underline{r})]^T P_c [A + \Delta A(\underline{r})] - P_c\} \underline{x} +$$

$$+ 2\underline{x}^T [A + \Delta A(\underline{r})]^T P_c [B + \Delta B(\underline{s})] \underline{g}(\underline{x}) +$$

$$+ \underline{g}^T(\underline{x}) [B + \Delta B(\underline{s})]^T P_c [B + \Delta B(\underline{s})] \underline{g}(\underline{x})$$

$$\leq - \|\underline{x}\|^2 \qquad (5.33)$$

for all pairs $(\underline{x},k) \in R^n x \mathcal{Z}$.

<u>Remark 5.4.1</u>

If inequality (5.33) is satisfied, then it can be proven that given any admissible uncertainties $\underline{r}(.)$ and $\underline{s}(.)$, the corresponding closed-loop system has $\underline{x} = \underline{0}$ as an asymptotically stable equilibrium point [39,68]. In the remainder of the paper, however

$$\underline{u}(k) = \underline{g}(\hat{\underline{x}}(k)) \tag{5.34}$$

will be taken, where $\hat{\underline{x}}(k)$ is the state estimate generated by a Luenberger observer.

Definition 5.4.1

Given $\underline{g}(.):R^n \longrightarrow R^m$ as per assumption (A4), let $\underline{\phi}(.,.)$: $R^n x R^n \longrightarrow R^m$ be defined as

$$\underline{\phi}(\underline{x}_1, \underline{x}_2) = \underline{g}(\underline{x}_1 - \underline{x}_2) - \underline{g}(\underline{x}_1) \tag{5.35}$$

As a consequence of the assumed cone-boundedness of $\underline{g}(.)$ and hence $\underline{\phi}(.,.)$, there will exist constants $\delta \geq 0$, $a \in [0, 2\delta]$ and $b \in [0,\delta]$ such that

$$\| \underline{g}(\underline{x}_1) \| \leq \delta \| \underline{x}_1 \| \tag{5.36}$$

$$\| \underline{\phi}(\underline{x}_1, \underline{x}_2) \| \leq a \| \underline{x}_1 \| + b \| \underline{x}_2 \| \tag{5.37}$$

for all $(\underline{x}_1, \underline{x}_2) \in R^{2n}$

Remark 5.4.2

If $\underline{g}(.)$ satisfies a Lipschitz condition,then there exists a non-negative constant Γ, called a Lipschitz constant, such that

$$\| \underline{g}(\underline{x}_1) - \underline{g}(\underline{x}_2) \| \leq \Gamma \| \underline{x}_1 - \underline{x}_2 \| \tag{5.38}$$

for all $(\underline{x}_1, \underline{x}_2) \in R^{2n}$ and in this case, a in (5.37) can be chosen to be zero.

Remark 5.4.3

When the feedback control law is linear; i.e. $\underline{u}(k) = G\hat{\underline{x}}(k)$, then the Lipschitz constant can be chosen to be $\|G\|$ and a can again be chosen to be zero, while $\delta = b = \Gamma = \|G\|$. For more details, see [73,74].

5.5 Construction of Full-order Observer

Design of linear and/or nonlinear state feedback control for discrete-time uncertain dynamical systems is accomplished [68-70] under the assumption that all state variables can be used in forming feedback signals. For situations where the entire state feedback is not measured, but only output measurements are available, it is necessary to reconstruct the states from the outputs. This is possible in principle provided that (A3) is met [76-78].

It is worth mentioning that the problem of designing observers for discrete control systems is not directly analogous to the continuous case. The main reason is due to the delay between measuring and processing the information [79,80]. To this end, the system, defined by (5.31) and (5.32), is considered with the following additional assumption :

(A5). There is a one-step delay between measuring and processing the information required. That is, the observation records consists of the measurements $\{\underline{y}(k-1), \underline{y}(k-2), \ldots\ldots, \underline{y}(0)\}$.

Therefore, it is fairly straightforward to design a state-reconstructor of the form:

$$\underline{\hat{x}}(k+1) = A\,\underline{\hat{x}}(k) + B\,\underline{u}(k) + M\,\{\underline{y}(k) - D\underline{u}(k) - C\underline{\hat{x}}(k)\} \qquad (5.39)$$

where $\underline{\hat{x}}(k) \in R^n$ is the state estimate of $\underline{x}(k)$ and $M \in R^{n \times p}$ is the observer gain matrix which may be suitable selected to ensure any desired degree of convergence of the observation scheme. Representing the mismatch in the modeling process by the state reconstruction error $\underline{e}(k) \in R^n$ defined as :

$$\underline{e}(k) = \underline{x}(k) - \underline{\hat{x}}(k) \qquad (5.40)$$

Application of (4.40) to (4.34) and in view of (4.35), yields

$$\underline{u}(k) = \underline{g}(\underline{x}(k)) + \underline{\phi}(\underline{x}(k),\underline{e}(k)) \qquad (5.41)$$

In view of (5.32) and (5.41) , one can obtain the following state and error equations for the closed-loop uncertain system

$$\underline{x}(k+1) = [A + \Delta A(\underline{r})]\,\underline{x}(k) + [B + \Delta B(\underline{s})]\,\underline{g}(\underline{x}(k)) +$$

$$+ [B + \Delta B(\underline{s})]\,\underline{\phi}(\underline{x}(k),\underline{e}(k)) \qquad (5.42)$$

$$\underline{e}(k+1) = A_e\,\underline{e}(k) + [\,\Delta A(\underline{r}) - M\,\Delta C(\underline{v})]\,\underline{x}(k) +$$

$$+ [\,\Delta B(\underline{s}) - M\,\Delta D(\underline{w})]\,\underline{g}(\underline{x}(k)-\underline{e}(k)) \qquad (5.43)$$

where

$$A_e = A - M\,C \qquad (5.44)$$

<u>Remark 5.5.1</u>

It has been pointed out in [44] that there exist quadratically stabilizable systems for which the stabilizing control law must be nonlinear. So, the choice of $\underline{g}(.)$ to be nonlinear is preferred.

Remark 5.5.2

Given an observer gain matrix M such that the eigenvalues $\lambda(A_e)$ have moduli strictly less than unity, it follows that there exists a unique nxn positivedefinite symmetric matrix P_o solving the Lyapunov equation

$$A_e^T P_o A_e - P_o = -Q_o \quad ; \quad Q_o = Q_o^T > 0 \tag{5.45}$$

5.6 Stability and Robustness Analysis

In this section, the stability and robustness properties of the overall system is studied. To do this, we construct a Lyapunov function defined as:

$$V(\underline{x},\underline{e},k) = \alpha_c \; \underline{x}^T(k) \; P_c \; \underline{x}(k) + \underline{e}^T(k) \; P_o \; \underline{e}(k) \tag{5.46}$$

where $\alpha_c > 0$ is a "tuning parameter" which will later be specified. Then, we will provide a set of sufficient conditions, whose satisfaction will ensure the asymptotic stability of the combined system. For ease of exposition, we define the following norm quantities :

$$\beta_1 = \max_{\mathcal{R}} \| [A + \Delta A(\underline{r})]^T \; P_c \; [A + \Delta A(\underline{r})] \| \tag{5.47a}$$

$$\beta_2 = \max_{\mathcal{R},\mathcal{S}} \| [B + \Delta B(\underline{s})]^T \; P_c \; [A + \Delta A(\underline{r})] \| \tag{5.47b}$$

$$\beta_3 = \max_{\mathcal{R},\mathcal{V}} \| A_e \; P_o \; [\Delta A(\underline{r}) - M \; \Delta C(\underline{v})] \| \tag{5.47c}$$

$$\beta_4 = \max_{\mathcal{S},\mathcal{W}} \| A_e \; P_o \; [\Delta B(\underline{s}) - M \; \Delta D(\underline{w})] \| \tag{5.47d}$$

$$\beta_5 = \max_{\mathcal{R},\mathcal{V}} \| [\Delta A(\underline{r}) - M \; \Delta C(\underline{v})]^T \; P_o \; [\Delta A(\underline{r}) - M \; \Delta C(\underline{v})] \| \tag{5.47e}$$

$$\beta_6 = \max_{\mathcal{R},\mathcal{S},\mathcal{V},\mathcal{W}} \| [\Delta A(\underline{r}) - M \; \Delta C(\underline{v})]^T \; P_o \; [\Delta B(\underline{s}) - M \; \Delta D(\underline{w})] \| \tag{5.47f}$$

$$\beta_7 = \max_{\mathcal{S},\mathcal{W}} \| [\Delta B(\underline{s}) - M\, \Delta D(\underline{w})]^T P_o\, [\Delta B(\underline{s}) - M\, \Delta D(\underline{w})] \| \qquad (5.47g)$$

<u>Theorem 5.6.1</u>

Consider the uncertain dynamical system (5.31) and (5.32) satisfying assumptions (A1)-(A5) with an observer-based control in the form described by (5.39) and (5.41). Moreover, suppose that the observer gain matrix M has been selected such that the eigenvalues of A_e have moduli strictly than unity. Then, the combined state and error system (5.42) and (5.43) is guranteed asymptotic stable with Lyapunov function (5.46) provided that the following sufficient conditions be satisfied

$$2\,(\beta_1 + \delta)\, a + \beta_2\, a^2 < 1 \qquad (5.48a)$$

$$\delta\,(2\,\beta_4 + \beta_7\,\delta) + \alpha_c\,\beta_2\, b^2 < \varsigma_o \qquad (5.48b)$$

$$\{\alpha_c\,[1 - 2\,(\beta_1 + \delta)\, a - \beta_2\, a^2] - [\beta_5 + 2\,\beta_6\,\delta + \beta_7\,\delta^2]\} \cdot$$

$$\cdot \{\varsigma_o - \delta\,(2\,\beta_4 + \beta_7\,\delta) - \alpha_c\,\beta_2\, b^2\} >$$

$$\{\alpha_c\,(\beta_1\, b + \delta\, b + \beta_2\, a\, b) + [\beta_3 + (\beta_4 + \beta_6)\,\delta + \beta_7\,\delta^2]\}^2 \qquad (5.48c)$$

where

$$\alpha_c > [\beta_5 + 2\beta_6\,\delta + \beta_7\,\delta^2]/[1 - 2(\beta_1 + \delta)a - \beta_2\, a^2] \qquad (5.49a)$$

and

$$\varsigma_o = \lambda_{min}(Q_o) \qquad (5.49b)$$

Proof

Consider a candidate Lyapunov function in (5.46) where $\alpha_c > 0$ is a tuning parameter, P_c is as per assumption A4 and P_o is the solution of (5.45). Then, given any admissible uncertainties $\underline{r}(.)$, $\underline{s}(.)$, $\underline{v}(.)$ and $\underline{w}(.)$, it follows that the Lyapunov forward difference corresponding to the combined state-error system (5.42) and (5.43) is given by :

$$\Delta V(k) = V(\underline{x}(k+1),\underline{e}(k+1)) - V(\underline{x}(k),\underline{e}(k))$$

$$= \alpha_c\ \underline{x}^T(k+1)\ P_c\ \underline{x}(k+1) + \underline{e}^T(k+1)\ P_o\ \underline{e}(k+1) -$$

$$- \alpha_c\ \underline{x}^T(k)\ P_c\ \underline{x}(k) - \underline{e}^T(k)\ P_o\ \underline{e}(k) \qquad (5.50)$$

Recalling (5.41)-(5.43) and dropping the suffices for simplicity, (5.50) can be written as

$$\Delta V = \alpha_c\ \{[A + \Delta A(\underline{r})]\underline{x} + [B + \Delta B(\underline{s})]\ [\underline{g}(\underline{x}) + \underline{\phi}(\underline{x},\underline{e})\}^T\ P_c \cdot$$

$$\cdot\ \{[A + \Delta A(\underline{r})\}\underline{x} + [B + \Delta B(\underline{s})]\ [\underline{g}(\underline{x}) + \underline{\phi}(\underline{x},\underline{e})\} +$$

$$+ \{A_e\ \underline{e} + [\Delta A(\underline{r}) - M\ \Delta C(\underline{v})]x + [\Delta B(\underline{s}) - M\ \Delta D(\underline{w})]\underline{g}(\underline{x}-\underline{e})\}^T \cdot$$

$$\cdot P_o\ \{A_e\ \underline{e} + [\Delta A(\underline{r}) - M\ \Delta C(\underline{v})]x + [\Delta B(\underline{s}) - M\ \Delta C(\underline{v})]\underline{g}(\underline{x}-\underline{e})\}$$

$$- \alpha_c\ \underline{x}^T\ P_c\ \underline{x} - \underline{e}^T\ P_o\ \underline{e} \qquad (5.51)$$

A little algebra on (5.51) using (5.33),(5.36)-(5.38), (5.45), (5.47) and (5.49b), yields

$$\Delta V_{max} \le -\alpha_c\ \|\underline{x}\|^2 + 2\alpha_c\ \beta_1\ \|\underline{x}\|\ (a\ \|\underline{x}\| + b\ \|\underline{e}\|) +$$

$$+ \alpha_c\ \beta_2\ (a\ \|\underline{x}\| - b\ \|\underline{e}\|)^2 - \rho_o\ \|\underline{e}\|^2 + 2\ \beta_3\ \|\underline{x}\|\cdot\|\underline{e}\| +$$

$$+ 2 \beta_4 \delta \|\underline{e}\| (\|\underline{x}\| + \|\underline{e}\|) + \beta_5 \|\underline{x}\|^2 + 2 \beta_6 \delta \|\underline{x}\| \cdot$$

$$\cdot (\|\underline{x}\| + \|\underline{e}\|) + \beta_7 \delta^2 (\|\underline{x}\| + \|\underline{e}\|)^2 \quad (5.52)$$

which in turn can be put in the form

$$\Delta V_{max} \leq - \underline{Z}^T \Omega \underline{Z} \quad (5.53)$$

where $\underline{Z} = (\|\underline{x}\| \quad \|\underline{e}\|)^T$ (5.54)

and Ω is a 2x2 matrix having entries

$$\Omega_{11} \triangleq \alpha_c [1 - 2 (\beta_1 + \delta) a - \beta_2 a^2] - [\beta_5 + 2 \beta_6 \delta + \beta_7 \delta^2] \quad (5.55a)$$

$$- \Omega_{12} = - \Omega_{21} \triangleq \alpha_c [\beta_1 b + \delta b + \beta_2 a b] + + [\beta_3 + (\beta_4 + \beta_6)\delta + \beta_7 \delta^2] \quad (5.55b)$$

$$\Omega_{22} \triangleq \rho_o - \delta (2 \beta_4 + \beta_7 \delta) - \alpha_c \beta_2 b^2 \quad (5.55c)$$

Now, it is clear that the overall system is guaranteed asymptotic stability if Ω is positive-definite matrix, i.e.

$$\Omega_{11} > 0 \quad ; \quad |\Omega| \triangleq \Omega_{11} . \Omega_{22} - \Omega_{12}^2 > 0 \quad (5.56)$$

which directly give the conditions (5.48), (5.49). Now, if given any admissible uncertainties $\underline{r}(.)$, $\underline{s}(.)$, $\underline{v}(.)$ and $\underline{w}(.)$ with α_c specified by (27) and the conditions (5.48), (5.49a) hold, we can conclude that

$$\Delta V_{max}(\underline{x}, \underline{e}, k) \leq - \lambda_{min}(\Omega) (\|\underline{x}\|^2 + \|\underline{e}\|^2) \quad (5.57)$$

for all $(\underline{x}, \underline{e}, k) \in R^n \times R^n \times \mathcal{Z}$.

This completes the proof of guaranteed asymptotic stability.

■■■

To this end, we have the following important results as special cases of the Theorem 5.6.1.

<u>Corollary 5.6.1</u>

If the state feedback control is linear, that is

$$\underline{u}(k) = G\,\hat{\underline{x}}(k) = G\,[\underline{x}(k) - \underline{e}(k)] \tag{5.58}$$

then, in view of Remark 5.4.3, it follows that the conditions (5.48),(5.49) are simplified to

$$[2\beta_4 + \beta_7\|G\|]\,\|G\| < \varrho_0 \tag{5.59a}$$

$$[\alpha_c - (\beta_5 + 2\beta_6\|G\| + \beta_7\|G\|^2)].[\varrho_0 - 2\beta_4\|G\| - \beta_7\|G\|^2 - \alpha_c\beta_2\|G\|^2] > \{\alpha_c(\beta_1\|G\| + \|G\|^2) + [\beta_3 + (\beta_4 + \beta_6)\|G\| + \beta_7\|G\|^2\}^2 \tag{5.59b}$$

$$\alpha_c > \beta_5 + 2\beta_6\|G\| + \beta_7\|G\|^2 \tag{5.59c}$$

<u>Corollary 5.6.2</u>

If the state feedback controller is non-linear but the uncertainties are vanished, i.e.

$$\Delta A(\underline{r}) = \Delta B(\underline{s}) = \Delta C(\underline{v}) = \Delta D(\underline{w}) = 0 \tag{5.60}$$

then, the conditions (5.48),(5.49) become :

$$2(\tilde{\beta}_1 + \delta)a + \tilde{\beta}_2 a^2 < 1 \qquad (5.61a)$$

$$\alpha_c \tilde{\beta}_2 b^2 < \rho_o \qquad (5.61b)$$

$$\alpha_c[1 - 2(\tilde{\beta}_1 + \delta) - \tilde{\beta}_2 a^2].(\rho_o - \alpha_c \tilde{\beta}_2 b^2) >$$

$$\alpha_c b^2 [\tilde{\beta}_1 + \delta + \beta_2 a]^2 \qquad (5.61c)$$

where $\alpha_c > 0$ and $\tilde{\beta}_1$, $\tilde{\beta}_2$ are defined as

$$\tilde{\beta}_1 = \| A^T P_c A\| \; ; \; \tilde{\beta}_2 = \| B^T P_c B\| \qquad (5.62)$$

Corollary 5.6.3

If the state feedback control is linear and the system (5.31) , (5.32) is completely deterministic , then the conditions (5.48), (5.49) reduced to a single one,given by

$$\rho_o > \alpha_c \|G\|^2 [\tilde{\beta}_2 + (\tilde{\beta}_1 + \|G\|)^2] \qquad (5.63)$$

where $\alpha_c > 0$ and $\tilde{\beta}_1$, $\tilde{\beta}_2$ are as in (5.62).

5.7 Example and Concluding Remarks

In this section, we apply our observer-based controller design to a typical system. We consider the positioning system [76] whose standard open-loop version is given by :

$$\dot{\underline{x}}(t) = \begin{bmatrix} 0 & 1 \\ 0 & -F/J \end{bmatrix} \underline{x}(t) + \begin{bmatrix} 0 \\ K/J \end{bmatrix} u(t)$$

where the state vector has components x_1 and x_2 representing the angular displacement and the angular velocity , while the control input u(t) is the input voltage to the motor. The choice of parameter values F = 46 Nm.rad.sec., J = 10 Kg.m², K = 7.87 Nm.rad/volt, together with a discretization scheme with time increment $\Delta t = 0.1$ sec., and with the parameters F/J and K/J undergo ± 10 % variation about their nominal values, yields the matrices of the sys tem (5.31), (5.32) of the form:

$$A = \begin{bmatrix} 1 & 0.08015 \\ 0 & 0.6313 \end{bmatrix} ; \quad B = \begin{bmatrix} 0.003396 \\ 0.06308 \end{bmatrix} ;$$

$$C = (1 \quad 0) \quad ; \quad D = [0] ;$$

$$\Delta A(\underline{r}) = \begin{bmatrix} 0 & r_1 \\ 0 & r_2 \end{bmatrix} ; \quad \Delta B(\underline{s}) = \begin{bmatrix} s_1 \\ s_2 \end{bmatrix} ;$$

$$\Delta C(\underline{v}) = \Delta D(\underline{w}) = 0 ;$$

and the compact bounding sets $\mathcal{R}$, $\mathcal{S}$ are given by :

$$\mathcal{R} = \{ \underline{r} \in R^2 : |r_1| \leq .0017, |r_2| \leq .03 \}$$

$$\mathcal{S} = \{ \underline{s} \in R^2 : |s_1| \leq .0003, |s_2| \leq .005 \}$$

Now, using the theory developed in [69], we obtain the control law in the form:

$$u(k) = -(110.4 \quad 12.66)\, \hat{\underline{x}}(k) + \underline{g}(\hat{\underline{x}}(k))$$

where g(.) is given by

$$g(\hat{\underline{x}}(k)) = \begin{cases} -\sigma(\hat{\underline{x}}).\theta(\hat{\underline{x}})/|\theta(\hat{\underline{x}})| & \text{for } |\theta(\hat{\underline{x}})| > \epsilon \\ -\sigma(\hat{\underline{x}}).\theta(\hat{\underline{x}})/\epsilon & \text{for } |\theta(\hat{\underline{x}})| \le \epsilon \end{cases} ;$$

$$\sigma(\hat{\underline{x}}) = 1.086 \{ |0.476\, \hat{x}_2| + |8.755\, \hat{x}_1 + 1.004\, \hat{x}_2| \} ;$$

$$\theta(\hat{\underline{x}}) = -(\,.31285 \quad .00042\,)\, \hat{\underline{x}} ;$$

and ϵ is a prespecified positive constant, chosen by the designer (here, we took $\epsilon = 0.5$).

Moreover, we select the observer gain matrix to be

$$M = (1.159 \quad 7.143)^T$$

For simulation purposes, the uncertain parameters $\underline{r}(k)$ and $\underline{s}(k)$ were taken to be sinusoidal functions of time in the form:

$$r_1(k) = 0.0017 \sin(0.05k) \quad ; \quad r_2(k) = 0.02 \sin(0.25k) ;$$

$$s_1(k) = 0.0003 \sin(0.15k) \quad \text{and} \quad s_2(k) = 0.005 \sin(0.2k).$$

The overall system dynamics were simulated from the initial conditions

$$\underline{x}(0) = (0.1 \quad 0)^T \quad , \quad \hat{\underline{x}}(0) = (0.2 \quad -0.1)^T$$

and the obtained closed-loop state trajectories are shown in Figs. 5.5 and 5.6. From these results, it can be concluded that the designed observer-based nonlinear control structure maintains the asymptotic stability of the overall system and the estimated states are very close to the true ones.

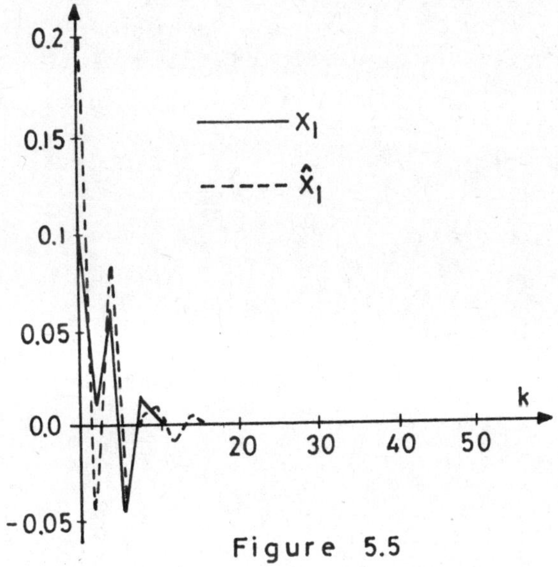

Figure 5.5

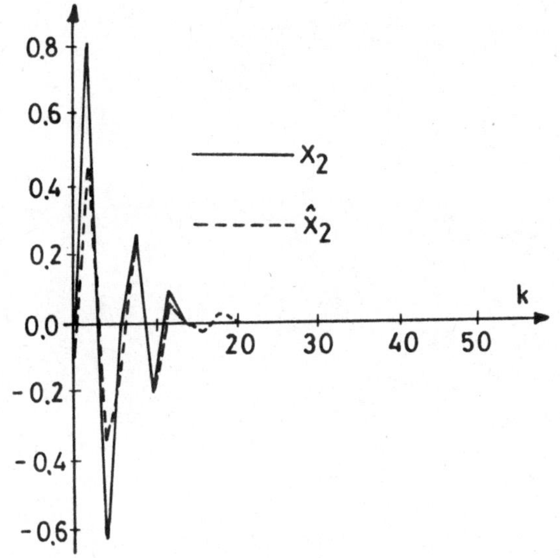

Figure 5.6

At this point, it should be emphasized that we considered a one-step delay in the measurement pattern. In the situation where the outputs are recorded instantaneously, it would be of interest to focus attention on the observers of current type; e.g. see [79,80]. Another area for future research would involve the optimal selection of the observer gain matrix M. Further work in these directions is in progress.

CHAPTER 6

INTERCONNECTED SYSTEMS : CONTINUOUS CASE

6.1 Introduction

During recent years, a number of papers have appeared which deal with the design of stabilizing controllers for uncertain dynamical systems [38-45]. In these papers, the uncertain parameters of the system are described only in terms of bounds on their possible sizes ; that is, statistical properties of the uncertain elements are not assumed to be known. Within this framework and using Lyapunov functions, controllers which guarantee stable operation for all possible variation quantities have been found. These controllers proved to guarantee uniform ultimate boundedness behaviour of the system under certain hypothesis on the bounded system uncertainties as well as on bounded admissible additive disturbances.

Large-scale interconnected dynamical systems are very susceptible to the uncertainties in the framework described above [87-90]. This arises from the fact that the interconnected systems' parameters cannot be calculated with sufficient accuracy to be used in on-line controllers. On the other hand, the stability of the present stabilizing control schemes is based on the assumption that all the information available about the system and the calculations based upon this information are centralized ; that is, take place at a single location. When considering large-scale systems the presupposition of centrality fails to hold due to either difficulties in simultaneous adjustment of a large number of parameters, or the difficulties in treating interconnection terms, or the lack of centralized information, or finally the lack of centralized computing capability.

In this chapter, we consider the design of decentralized and hierarchical control techniques in order to: (1) enhance stability and improve operating conditions of interconnected dynamical systems incorporating bounded uncertain elements, (2) deal with the inevitable unmodeled bounded uncertainties that may be present in the interconnections between the subsystems, (3) prevent or at least minimize the necessity of information exchanging between the decomposed subsystems , and (4) reduce the computational difficulties of the present solution algorithms which may lead to computational difficulties and/or numerical instability. It has been shown that our methodologies can lead to stabilizable control strategies which maintain the system stability and reliability under normal operations as well as some situations with failures in the interconnections and/or the data transmitting network.

6.2 Problem Formulation

Consider a class of uncertain dynamical systems shown in Fig. 6.1 and described as an interconnection of N subsystems, i.e.

$$\dot{\underline{x}}_i = [A_i + \Delta A_i(\underline{r}_i(t))]\, \underline{x}_i(t) + [B_i + \Delta B_i(\underline{s}_i(t))]\, \underline{u}_i(t)$$

$$+ C_i\, \underline{v}_i(t) + \underline{Z}_i + \underline{h}_i \quad ; \; i=1,2,\ldots,N \tag{6.1a}$$

$$\underline{Z}_i = \sum_{\substack{j=1 \\ j\neq i}}^{N} A_{ij}\, \underline{x}_j \tag{6.1b}$$

$$\underline{h}_i = \sum_{\substack{j=1 \\ j\neq i}}^{N} H_{ij}(t, \underline{r}_i(t), \underline{x}_j(t)) \tag{6.1c}$$

$$\underline{x}_i(t_0) = \underline{x}_{i0} \tag{6.1d}$$

$$\underline{y}_i(t) = \underline{x}_i(t) + \underline{w}_i(t) \tag{6.2}$$

where for the ith subsystem : $\underline{x}_i \in R^{ni}$ is the state , $\underline{u}_i \in R^{mi}$ is the control , $\underline{v}_i \in R^{li}$ is bounded disturbance , $\underline{y}_i \in R^{ni}$ is the measured state , $\underline{w}_i \in R^{ni}$ is the measurement error , $Z_i \in R^{ni}$ is the interconnection with the other subsystems and the constant matrices A_i , B_i , C_i , A_{ij} are prescribed with appropriate dimensions. The uncertainty parameters are $\underline{r}_i \in \mathcal{R}_i \subset R^{pi}$, $\underline{s}_i \in \mathcal{S}_i \subset R^{qi}$, $\underline{v}_i \in \mathcal{V}_i \subset R^{li}$ and $\underline{w}_i \in \mathcal{W}_i \subset R^{ni}$. The subsystem matrix uncertainty $\Delta A_i(\underline{r}_i)$ and the input component matrix uncertainty $\Delta B_i(\underline{s}_i)$ depend on parameters $\underline{r}_i$ and $\underline{s}_i$ respectively. The term $C_i \; \underline{v}_i(t)$ accounts for external input uncertainty. Moreover , $\underline{h}_i \in R^{ni}$ contains the uncertainties in the interconnections between the ith subsystem and the other subsystems.

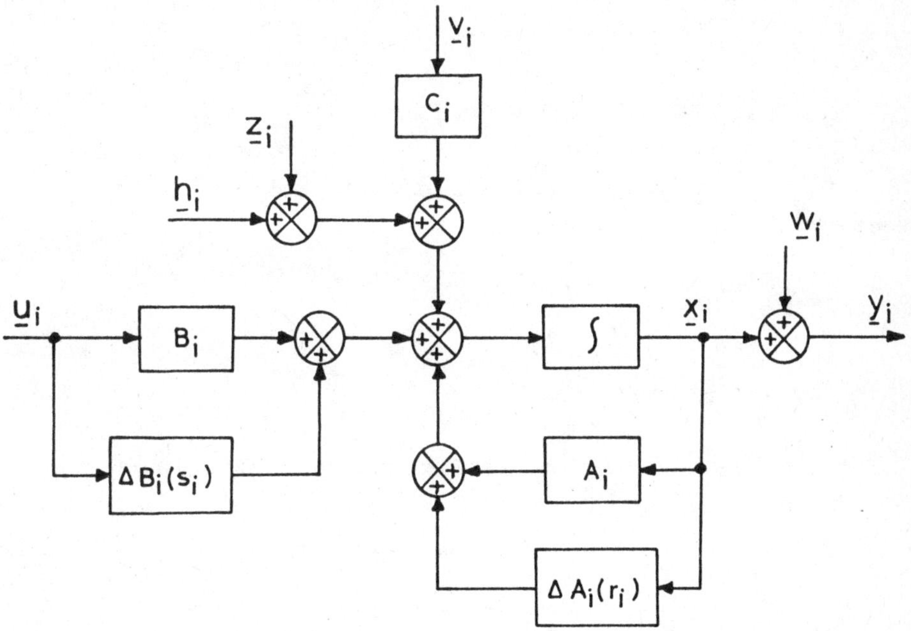

Figure 6.1: Additive-type Uncertain Subsystem.

Throughout the remainder of this Chapter, the following assumptions are taken for each subsystem as standard :

<u>Assumption 1 (continuity) :</u> The matrix functions $\Delta A_i(.)$ and $\Delta B_i(.)$ depend continuously on their arguments.

<u>Assumption 2 (compactness) :</u> The bounding sets $\mathcal{R}_i$, $\mathcal{S}_i$, $\mathcal{V}_i$ and $\mathcal{W}_i$ are compact sets in the indicated finite-dimensional Euclidean spaces.

<u>Assumption 3 (measurability) :</u> The uncertainties $\underline{r}_i(.):[0,\infty) \longrightarrow \mathcal{R}_i$, $\underline{s}_i(.):[0,\infty) \longrightarrow \mathcal{S}_i$, $\underline{v}_i(.):[0,\infty) \longrightarrow \mathcal{V}_i$, and $\underline{w}_i(.):[0,\infty) \longrightarrow \mathcal{W}_i$ are assumed to be Lebesgue measurable.

<u>Assumption 4 (stabilizability) :</u> The pair (A_i, B_i) is stabilizable ; that is there exists, a constant $(m_i \times n_i)$ matrix G_i , such that the eigenvalues $\lambda(\tilde{A}_i) \triangleq \lambda(A_i + B_i G_i)$ are strictly in the complex left half plane.

<u>Assumption 5 (matching conditions) :</u>

(i) There exist matrix functions (of appropriate dimensions) $D_i(.)$ and $E_i(.)$ whose entries are continuous on R^p and R^q respectively such that :

$$\Delta A_i(\underline{r}_i) = B_i D_i(\underline{r}_i) \tag{6.3a}$$

$$\Delta B_i(\underline{s}_i) = B_i E_i(\underline{s}_i) \tag{6.3b}$$

(ii) There exists a constant matrix function F_i such that

$$C_i = B_i F_i \tag{6.3c}$$

(iii) $$\max_{s_i \in \mathcal{S}_i} \| E_i(\underline{s}_i) \| < 1 \tag{6.3d}$$

In terms of (6.1)-(6.3) the composite system can be

described as :

$$\dot{\underline{x}}(t) = [A + \Delta A(\underline{r}) + M]\,\underline{x}(t) + [B + \Delta B(\underline{s})]\,\underline{u}(t) + C\,\underline{v}(t) +$$

$$+ H(t,\underline{x}(t),\underline{r}(t)) \tag{6.4a}$$

$$\underline{x}(0) = \underline{x}_0 \tag{6.4b}$$

$$\underline{y}(t) = \underline{x}(t) + \underline{w}(t) \tag{6.5}$$

where $\underline{x} = [\underline{x}_1^T, \underline{x}_2^T, \ldots\ldots, \underline{x}_N^T]^T \in R^n$ is the composite state ; $n = \sum_{i=1}^{N} n_i$, $\underline{u} = [\underline{u}_1^T, \underline{u}_2^T, \ldots\ldots, \underline{u}_N^T]^T \in R^m$ is the composite control ; $m = \sum_{i=1}^{N} m_i$, $\underline{v} = [\underline{v}_1^T, \underline{v}_2^T, \ldots\ldots, \underline{v}_N^T]^T \in R^l$ is the composite disturbance ; $l = \sum_{i=1}^{N} l_i$, $\underline{y} = [\underline{y}_1^T, \underline{y}_2^T, \ldots\ldots, \underline{y}_N^T]^T$ $\in R^n$ is the composite measured state , $\underline{w} = [\underline{w}_1^T, \underline{w}_2^T, \ldots, \underline{w}_N^T]^T$ $\in R^n$ is the composite measurement error , M and $H(.,.,.)$ represent respectively the interactions and their uncertainties of the overall system. Furthermore , $A = \text{diag}\,(A_i)$, $B = \text{diag}\,(B_i)$, $C = \text{diag}\,(C_i)$, $\Delta A(.) = \text{diag}\,(\Delta A_i(.))$ and $\Delta B(.) = \text{diag}\,(\Delta B_i(.))$. It is interesting to note that both M and H are off-diagonal matrices , i.e. $M_{ij} = H_{ij} = 0$ for $i \neq j$.

Our goal is to design decentralized (local) and two-level (hierarchical) control strategies in order to stabilize a class of interconnected uncertain dynamic systems perturbed with uncertainties in the framework described above.

Before going into the analysis , we have to recall preliminary results of ultimate boundedness theory [38-45] :

Definition 6.2.1 (Practical Stabilizability)

The uncertain composite dynamical system (6.4) is said to be practically stabilizable if , given any $\underline{d} > 0$, there is a control law $\underline{g}(.,.) : R^n \times R \longrightarrow R^m$ for which , given any admissible uncertainties $\underline{r}(.) \in \mathcal{R}$, $\underline{s}(.) \in \mathcal{S}$, $\underline{v}(.) \in \mathcal{V}$ and $\underline{w}(.) \in \mathcal{W}$ (where $\mathcal{R} = \bigcup_{i=1}^{N} \mathcal{R}_i$, $\mathcal{S} = \bigcup_{i=1}^{N} \mathcal{S}_i$ and so on), any initial time $t_0 \in R$ and any initial state $\underline{x}_0 \in R^n$, the following conditions hold :

(i) The closed loop system

$$\dot{\underline{x}}(t) = [A + \Delta A(\underline{r}) + M]\, \underline{x}(t) + [B + \Delta B(\underline{s})]\, \underline{g}(\underline{y}(t),t) + C\, \underline{v}(t) + \\ + H(t,\underline{r}(t),\underline{x}(t)) \qquad (6.6)$$

possesses a solution $\underline{x}(.) : [t_0, t_1] \longrightarrow R^n$, $x(t_0) = x_0$

(ii) Given any $\nu > 0$ and any solution $x(.) : [t_0, t_1] \longrightarrow R^n$, $\underline{x}(t_0) = \underline{x}_0$ of (6.6) with $\| \underline{x}_0 \| \le \nu$, there is a constant $d(\nu) > 0$ such that $\| \underline{x}(t) \| \le d(\nu)$ $\forall$ $t \in [t_0, t_1]$.

(iii) Every solution $\underline{x}(.) : [t_0, t_1] \longrightarrow R^n$ can be continued over $[t_0, \infty)$.

(iv) Given any $\bar{d} \ge \underline{d}$, any $\nu > 0$ and any solution $\underline{x}(.) : [t_0, \infty) \longrightarrow R^n$, $\underline{x}(t_0) = \underline{x}_0$ of (6.6) with $\| \underline{x}_0 \| \le \nu$, there exists a finite time $T(\bar{d},\nu) < \infty$, possibly dependent on ν but not on t_0 , such that $\| \underline{x}(t) \| \le \bar{d}$ $\forall$ $t \ge t_0 + T(\bar{d},\nu)$.

(v) Given any $\bar{d} \ge \underline{d}$ and any solution $\underline{x}(.) : [t_0, t_1] \longrightarrow R^n$, $\underline{x}(t_0) = \underline{x}_0$, of (6.6), there is a constant $\delta(\bar{d}) > 0$ such that $\| \underline{x}(t_0) \| \le \delta(\bar{d})$ implies $\| \underline{x}(t) \| \le \bar{d}$ $\forall$ $t \ge t_0$.

6.3 Decentralized Control Analysis of Decoupled Subsystems

By adopting the concept of decentralized control, the system is controlled using local control stations each is connected with a subsystem and receives only information concerning the local state variables. All these controllers are involved in controlling the overall interconnected system.

Now, given a matrix G_i satisfying Assumption 4, consider the class of decentralized feedback controls :

$$\underline{u}_i(t) = G_i \ \underline{y}_i(t) + \underline{g}_i(\underline{y}_i(t),t) \qquad \forall \quad \underline{y}_i \in R^n \tag{6.7}$$

where $g_i(.,.) : R^n \longrightarrow R^m$ is given by

$$\underline{g}_i(\underline{y}_i,t) = \begin{cases} \dfrac{-B_i^T P_i \underline{y}_i}{\| B_i^T P_i \underline{y}_i \|} \sigma_i(\underline{y}_i) & \text{for } \|B_i^T P_i \underline{y}_i\| > \epsilon_i \\[2ex] \dfrac{-B_i^T P_i \underline{y}_i}{\epsilon_i} \sigma_i(\underline{y}_i) & \text{for } \|B_i^T P_i \underline{y}_i\| \le \epsilon_i \end{cases} \tag{6.8}$$

where ϵ_i is a pre-specified positive constant, P_i is a solution of Lyapunov equation :

$$P_i \tilde{A}_i + \tilde{A}_i^T P_i = - Q_i \qquad ; \quad Q_i > 0 \tag{6.9}$$

and $\sigma_i(.) : R^{n_i} \longrightarrow R_+$ is non-negative function chosen to satisfy :

$$\sigma_i(\underline{y}_i) = [\ 1 - \max_{\underline{s}_i \in S_i} \|E_i(\underline{s}_i)\|\ ]^{-1} \{ \max_{\underline{r}_i \in \mathcal{R}_i} \|D_i(\underline{r}_i)\ \underline{y}_i\| +$$

$$+ \max_{\substack{\underline{r}_i \in \mathcal{R}_i \\ \underline{w}_i \in \mathcal{W}_i}} \| D_i(\underline{r}_i)\, \underline{w}_i \| + \max_{\underline{w}_i \in \mathcal{W}_i} \| G_i\, \underline{w}_i \|$$

$$+ \max_{\underline{s}_i \in \mathcal{S}_i} \| E_i(\underline{s}_i)\, G_i\, \underline{y}_i \| + \max_{\underline{v}_i \in \mathcal{V}_i} \| F_i\, \underline{v}_i \| \} \qquad (6.10)$$

For ease of exposition , we define the following norm quantities :

$$\rho_{ri} = \max_{\underline{r}_i \in \mathcal{R}_i} \| D_i(\underline{r}_i) \| \quad ; \quad \rho_{si} = \max_{\underline{s}_i \in \mathcal{S}_i} \| E_i(\underline{s}_i) \| \qquad (6.11a)$$

$$\rho_{vi} = \max_{\underline{v}_i \in \mathcal{V}_i} \| F_i\, \underline{v}_i \| \quad ; \quad \rho_{wi} = \max_{\underline{w}_i \in \mathcal{W}_i} \| \underline{w}_i(t) \| \qquad (6.11b)$$

$$\rho_{gi} = \| G_i \| \quad ; \quad \rho_{sgi} = \max_{\underline{s}_i \in \mathcal{S}_i} \| E_i(\underline{s}_i)\, G_i \| \qquad (6.11c)$$

Assumption 6 (cone boundedness) :

The functions $\sigma_i(\underline{y}_i)$, $i=1,2,\ldots,N$ are assumed to be cone-bounded , that is

$$\| \sigma_i(\underline{y}_i) \| \leq \alpha_i + \beta_i \| \underline{y}_i \| \qquad (6.12)$$

Assumption 7 (Interactions' boundedness) :

The interactions $H_{ij}(t, \underline{r}_i, \underline{x}_j)$ are assumed to be carathedory functions *. Furthermore , they are to be bounded , i.e.

---

* A function of $\underline{x}$ and t is a carathedory function if for all $t \in R$ it is continuous in $\underline{x}$ and and for all $\underline{x} \in R^n$, it is Lebesgue measurable in t.

$$\left\| \sum_{\substack{j=1 \\ \neq i}}^{N} H_{ij}(t, \underline{r}_i, \underline{x}_j) \right\| \leq \sum_{\substack{j=1 \\ \neq i}}^{N} \| \Delta A_{ij}(t, \underline{r}_i) \| . \| \underline{x}_j \|$$

$$\leq \sum_{\substack{j=1 \\ \neq i}}^{N} \gamma_{ij} \| \underline{x}_j \| \quad \forall \quad (t, \underline{r}_i, \underline{x}_j) \in R \times R^{p_i} \times R^{n_i} \; ;$$

$$; \quad i=1,2,\ldots\ldots, N \tag{6.13}$$

where γ_{ij} are N^2 non-negative upper bounds for the uncertainties among the subsystems.

REMARK 6.3.1 :

It should be emphasized that Assumption 6 is always satisfied because in view of (6.2), (6.11)-(6.12) , one can deduce that :

$$\sigma_i(\underline{y}_i) \leq \sigma_i(\underline{x}_i)$$

$$\triangleq a_i + b_i \| \underline{x}_i \| \tag{6.14}$$

where

$$a_i = [(2 \rho_{ri} + \rho_{gi} + \rho_{sgi}) \rho_{wi} + \rho_{vi}] / (1 - \rho_{si}) \tag{6.15a}$$

$$b_i = (\rho_{ri} + \rho_{sgi}) / (1 - \rho_{si}) \tag{6.15b}$$

6.4 Decentralized Control Stabilization

The following theorem sets up the condition for decentralized stabilization with respect to the strength (the magnitude of information flow) of the interactions as well as the interactions' uncertainties among various subsystems.

Theorem 6.4.1

The composite system (6.4) satisfying Assumptions (1-7) can be practically stabilized in a decentralized fashion by the local controls (6.7) if the test matrix given by $L = [l_{ij}]$

$$l_{ij} = \begin{cases} \lambda_m (Q_i) & i = j \\ - \lambda_M (P_i) [\| A_{ij} \| + \| A_{ji} \| + 2\gamma] & i \neq j \end{cases} \tag{6.16}$$

is positive definite matrix.
where γ , represents the total bound for interactions uncertainty , is defined as :

$$\gamma = \sum_{i=1}^{N} \sum_{j=1}^{N} \gamma_{ij} \tag{6.17}$$

Proof

The system (6.1) under the application of the control (6.7) and utilizing the Assumptions (1-7) can be written as :

$$\dot{\underline{x}}_i = A_i \underline{x}_i + B_i \underline{g}_i (\underline{x}_i + \underline{w}_i , t) + B_i \underline{\Phi}_i (\underline{x}_i , t) + \sum_{\substack{j=1 \\ j \neq i}}^{N} A_{ij} \underline{x}_j + \sum_{\substack{j=1 \\ j \neq i}}^{N} H_{ij} (t, \underline{r}_i , \underline{x}_j) \tag{6.18}$$

where

$$\underline{\Phi}_i (\underline{x}_i , t) = D_i (\underline{r}_i) \underline{x}_i + G_i \underline{w}_i + E_i (\underline{s}_i) G_i (\underline{x}_i + \underline{w}_i) + E_i (\underline{s}_i) \underline{g}_i (\underline{x}_i + \underline{w}_i , t) + F_i \underline{v}_i \tag{6.19}$$

In view of (6.8) and (6.10), one can easily show that :

$$\| \underline{\Phi}_i (\underline{x}_i, t) \| \le \sigma_i (\underline{y}_i) \tag{6.20}$$

Now , let the Lyapunov function V_i be as an index of " energy " associated with x_i and is chosen as

$$V_i = \underline{x}_i^T P_i \underline{x}_i \quad ; \quad i=1,2,\ldots\ldots,N \tag{6.21}$$

where P_i is the solution of (6.9). Then it is tempting to define

$$V = \sum_{i=1}^{N} V_i \tag{6.22}$$

as the energy of the composite system. Taking the rate of change of (6.21) , one gets

$$\dot{V}_i = \dot{\underline{x}}_i^T P_i \underline{x}_i + \underline{x}_i^T P_i \dot{\underline{x}}_i \tag{6.23}$$

A little algebra on (6.23) using (6.2), (6.8), (6.9), (6.20) and dropping the suffices for simplicity , yields :

$$\dot{V}_i = \underline{x}_i^T [P_i \tilde{A}_i + \tilde{A}_i^T P_i] \underline{x}_i + 2 \underline{x}_i^T P_i B_i [\underline{g}_i (\underline{y}_i, t) + \underline{\Phi}_i (\underline{x}_i, t)] +$$

$$+ 2 \underline{x}_i^T P_i \sum_{j=1}^{N} A_{ij} \underline{x}_j + 2 \underline{x}_i^T P_i \sum_{j=1}^{N} H_{ij} (t, \underline{r}_i, \underline{x}_j)$$

$$\le - \underline{x}_i^T Q_i \underline{x}_i + 2 (B_i^T P_i \underline{y}_i)^T [\underline{g}_i (\underline{y}_i, t) +$$

$$+ \sigma_i (\underline{y}_i) B_i^T P_i \underline{y}_i / \| B_i^T P_i \underline{y}_i \|] - 2 (B_i^T P_i \underline{w}_i)^T [\underline{g}_i (\underline{y}_i, t) -$$

$$- \sigma_i (\underline{y}_i) B_i^T P_i \underline{w}_i / \| B_i^T P_i \underline{w}_i \|] + 2 \underline{x}_i^T P_i \sum_{j=1}^{N} A_{ij} \underline{x}_j +$$

$$+ 2 \underline{x}_i^T P_i \sum_{j=1}^{N} H_{ij} (t, \underline{r}_i, \underline{x}_j) \tag{6.24}$$

As a consequence of (6.8), the second term on the r.h.s. of (6.24) vanishes for $\| B_i^T P_i \underline{y}_i \| > \epsilon_i$, but if $\| B_i^T P_i \underline{y}_i \| \leq \epsilon_i$, its maximum value (at $\| B_i^T P_i \underline{y}_i \| = \epsilon_i / 2$) is equal to $\epsilon_i \sigma_i(\underline{y}_i) / 2$. The maximum value of the third term occurring for $\| B_i^T P_i \underline{y}_i \| > \epsilon_i$ and $B_i^T P_i \underline{y}_i / \| B_i^T P_i \underline{y}_i \| = B_i^T P_i \underline{w}_i / \| B_i^T P_i \underline{w}_i \|$ is simply $4 \| B_i^T P_i \underline{w}_i \| \sigma_i(\underline{y}_i)$. The maximum value of the fourth term is equal to

$2 \lambda_M (P_i) \| \underline{x}_i \| \sum_{j=1}^{N} \| A_{ij} \| . \| \underline{x}_j \|$ and in view of (6.10), the maximum value of the last term is equal to

$2 \lambda_M (P_i) \| \underline{x}_i \| \sum_{j=1}^{N} \gamma_{ij} \| \underline{x}_j \|$. Regrouping the terms and summing up the maxima, we have :

$$\dot{V}_i \leq - \underline{x}_i^T Q_i \underline{x}_i + [\epsilon_i /2 + 4 \| B_i^T P_i \underline{w}_i \|] \sigma_i(\underline{y}_i) + $$

$$+ 2 \lambda_M (P_i) \| \underline{x}_i \| \sum_{j=1}^{N} (\| A_{ij} \| + \gamma_{ij}) \| \underline{x}_j \|$$

$$\leq - \underline{x}^T L \underline{x} + \mu_1 \| \underline{x} \| + \mu_0 \qquad (6.25)$$

where the NxN symmetric matrix L is given by (6.16) and

$$\mu_0 = \sum_{i=1}^{N} a_i (\epsilon_i /2 + 4 \| B_i^T P_i \| \rho_{wi}) , \qquad (6.26a)$$

$$\mu_1 = \sum_{i=1}^{N} b_i (\epsilon_i /2 + 4 \| B_i^T P_i \| \rho_{wi}) \qquad (6.26b)$$

Now , $\dot{V}$ is negative definite if the test matrix L is positive definite for all $(\underline{x},t) \in \Omega^c(\eta) x R$ where $\Omega^c(\eta)$ is the complement of the set $\Omega(\eta)$, whereas $\Omega(\eta)$ is a closed sphere with radius η given by

$$\eta = \{ \mu_1 + \sqrt{\mu_1^2 + 4 \mu_0 \lambda_m(L)} \} / (2 \lambda_m(L)) \tag{6.27}$$

Therefore , in view of the theory of ultimate boundedness [40,41,45], it suffices to get $\underline{d} > 0$ in definition 6.2.1. We define it as the major axis of the smallest ellipsoid containing $\Omega(\eta)$, using the standard arguments in [40,41,45] , $\underline{d}$ will be given by :

$$\underline{d} = \eta \sqrt{\lambda_M(P) / \lambda_m(P)} \tag{6.28}$$

where

$$P = \text{diag}(P_i) \tag{6.29}$$

This concludes the proof of Theorem 6.4.1. ■■■

Some observations are in order with regard to the foregoing analysis :

(1) An important selection of the linear part of the controller , namely G_i , would be the optimal feedback gain for each subsystem in order to ensure the stabilizability of A_i [94].

(2) One aspect of the attractiveness of the result of the Theorem 6.4.1 stems from the saving in computational effort in constructing the test matrix L as opposed to solving the Lyapunov equation directly for the overall system. The latter is of dimension nxn where $n = \sum_{i=1}^{N} n_i$, whereas L is of

dimension NxN and to construct L we need to solve N-separate Lyapunov equations of small dimensions which is much easier to solve.

(3) The decentralized controller structure (6.7) is easy to compute since we are working with a set of decomposed subsystems rather than the overall model. Moreover , it is physically realizable and easy to implement as we don't need to exchange any state information between the subsystems.

Corollary 6.4.1

In the absence of state uncertainty ; that is , $\underline{w}_i = \underline{0}$, and if there are infinite feedback gains ; that is , $\epsilon_i = 0$,then the nonlinear (switching) term of the controller (6.7) will be given by :

$$\underline{g}_i(\underline{y}_i, t) = \underline{g}_i(\underline{x}_i, t) =$$

$$= \begin{cases} -\sigma_i(\underline{x}_i)\,(B_i^T P_i \underline{x}_i) / \| B_i^T P_i \underline{x}_i \| & \text{for } \| B_i^T P_i \underline{x}_i \| \neq 0 \\ \epsilon_i \{ \underline{\xi}_i \in R^n : \| \underline{\xi}_i \| \leq \sigma_i(\underline{x}_i) & \text{for } \| B_i^T P_i \underline{x}_i \| = 0 \end{cases} \tag{6.30}$$

It follows from (6.27) in this case that $\eta = 0$ and Ω = empty set ; that is , Ω^c will represent the whole space. On the other hand from (6.25) , we find that

$$\dot{V} \leq -\underline{x}^T L \underline{x} \qquad \forall\ (\underline{x}, t) \in R^n \times R \tag{6.31}$$

6.5 Hierarchical Control Stabilizability

In sections 6.3 and 6.4 , we have seen that we can stabilize the overall uncertain system using decentralized controllers. This technique doesn't need any exchange of state information between the subsystems , and hence it is economic from the point of view of the control realization. However, in this completely decentralized control methodology, the interconnections between the subsystems are completely neglected, yielding a loss of information about the interconnections' behaviour and their dynamic effects on each subsystems. So in this section we investigate the stabilization problem via multilevel technique [85-92]. The principle idea of the hierarchical control theory [88] is to define a set of subproblems that can be considered independent at certain level (subsystem level). Through the manipulation of the interplaying effect at a higher level (coordinator), one obtains the global solution.

Now, the decentralized control (6.7) is modified to be in the hierarchical form :

$$\underline{u}(t) = G_b\ \underline{y}(t) + \underline{g}(\underline{y}(t),t) \tag{6.32}$$

where

$$G_b = \text{diag}\ (G_i) \quad ; \quad G_i \text{ satisfies Assumption 4} \tag{33}$$

and $\underline{g}(.,.) : R^n \longrightarrow R^m$ is given by

$$\underline{g}(\underline{y}(t),t) = \begin{cases} -\ \sigma(\underline{y})\ B^T\ P\ \underline{y}\ /\ \|B^T\ P\ \underline{y}\| & \text{for } \|B^T\ P\ \underline{y}\| > \epsilon \\ -\ \sigma(\underline{y})\ B^T\ P\ \underline{y}\ /\ \epsilon & \text{for } \|B^T\ P\ \underline{y}\| \leq \epsilon \end{cases} \tag{6.34}$$

where ϵ is a pre-specified positive constant, $P = \text{diag}\ (P_i)$; P_i is the solution of (6.9) and $\sigma(.) : R^n \longrightarrow R_+$ is non-negative function chosen to satisfy :

$$\sigma(\underline{y}) = \max_i \{\ [1 - \max_{\underline{s}_i \in \mathcal{S}_i} \|E_i(\underline{s}_i)\|]^{-1}\ [\ \max_{\underline{r}_i \in \mathcal{R}_i} \|D_i(\underline{r}_i)\ \underline{y}_i\| +$$

$$+ \underset{\substack{\underline{r}_i \in \mathcal{R}_i \\ \underline{w}_i \in \mathcal{W}_i}}{\text{Max}} \| D_i(\underline{r}_i)\, \underline{w}_i \| + \underset{\underline{w}_i \in \mathcal{W}_i}{\text{Max}} \| G_i\, \underline{w}_i \| +$$

$$+ \underset{\underline{s}_i \in \mathcal{S}_i}{\text{Max}} \| E_i(\underline{s}_i)\, G_i\, \underline{y}_i \| + \underset{\underline{v}_i \in \mathcal{V}_i}{\text{Max}} \| F_i\, \underline{v}_i \| \} \quad ; \quad i=1,2,\ldots\ldots,N \tag{6.35}$$

In view of (6.11), (6.12) and (6.14), we can find that :

$$\sigma(\underline{y}) \le \sigma(\underline{x}) \triangleq \tilde{a} + \tilde{b}\, \|\underline{x}\| \tag{6.36}$$

where

$$\tilde{a} \triangleq \underset{i}{\text{Max}} \{[(2\, \rho_{ri} + \rho_{gi} + \rho_{sgi})\, \rho_{wi} + \rho_{vi}]/(1 - \rho_{si})\} \tag{6.37a}$$

$$\tilde{b} \triangleq \underset{i}{\text{Max}} \{[\rho_{ri} + \rho_{sgi}]/(1 - \rho_{si})\} \tag{6.37b}$$

where $i=1,2,\ldots\ldots,N$

Theorem 6.5.1

The composite system (6.4) satisfying Assumptions (1-7) can be practically stabilized via the hierarchical control (6.32) if the test matrix L given by (6.16) is positive definite. Furthermore , the resulting closed - loop state trajectories are bounded in a set $\Omega^c(\bar{\eta})$ where $\Omega^c(\bar{\eta})$ is the complement of a set $\Omega(\bar{\eta})$ with radius $\bar{\eta}$ given by

$$\bar{\eta} = [\, \bar{\mu}_1 + \sqrt{\bar{\mu}_1^2 + 4\, \bar{\mu}_0\, \lambda_m(L)}\,] \,/\, [2\, \lambda_m(L)] \tag{6.38}$$

where

$$\bar{\mu}_0 = \tilde{a} \, (\epsilon/2 + 4 \, \| B^T P \| \, \rho_w) \qquad (6.39a)$$

$$\bar{\mu}_1 = \tilde{b} \, (\epsilon/2 + 4 \, \| B^T P \| \, \rho_w) \qquad (6.39b)$$

$$\rho_w = \underset{i}{\text{Max}} \, \{ \rho_{wi} \} \quad ; \quad i=1,2,\ldots,N \qquad (6.39c)$$

and $\tilde{a}$, $\tilde{b}$ are given by (6.37).

Proof

The closed loop system (6.4) with the control law (6.32) becomes :

$$\dot{\underline{x}} = (A + M) \, \underline{x} + B \, [\underline{g}(\underline{y},t) + \underline{\Phi}(\underline{x},t)] + H(t,\underline{r},\underline{x}) \qquad (6.40)$$

where

$$\underline{\Phi}(\underline{x},t) = D(\underline{r}) \, \underline{x} + E(\underline{s}) \, G \, (\underline{x} + \underline{w}) + E(\underline{s}) \, \underline{g}(\underline{y},t) + F \, \underline{v} \qquad (6.41)$$

In view of (6.35) and (6.36) , one gets :

$$\begin{aligned} \| \underline{\Phi}(\underline{x},t) \| &\le \sigma(\underline{y}) \\ &\le \sigma(\underline{x}) \\ &= \tilde{a} + \tilde{b} \, \| \underline{x} \| \end{aligned} \qquad (6.42)$$

where a and b are given by (6.37). Now , defining a candidate Lyapunov function , V , as

$$V = \underline{x}^T \, P \, \underline{x} \qquad (6.43)$$

where $P = \text{diag} \, (P_i)$; P_i is defined before.

Taking the rate of change of V , we have :

$$\dot{V} = 2\ \underline{x}^T P\ \underline{x}$$

$$= 2\ \underline{x}^T P\ (\tilde{A} + M)\ \underline{x} + 2\ \underline{x}^T P\ B\ [\underline{g}(\underline{y},t) + \underline{\Phi}(\underline{x},t)] +$$

$$+ 2\ \underline{x}^T P\ H(t,\underline{r},\underline{x})$$

$$\leq -\ \underline{x}^T Q\ \underline{x} + 2\ \underline{x}^T P\ M\ \underline{x} + [\epsilon/2 + 4\ \| B^T P \|\ \rho_w]\ \sigma(\underline{y}) +$$

$$+ 2\ \underline{x}^T P\ H(t,\underline{r},\underline{x})$$

$$\leq -\ \underline{x}^T L\ \underline{x} + \bar{\mu}_1\ \| \underline{x} \| + \bar{\mu}_0 \qquad (6.44)$$

where $\bar{\mu}_0$,$\bar{\mu}_1$ and ρ_w are given by (6.39).

As a result , $\dot{V}$ is negative definite if the requirements in the Theorem 6.5.1 be satisfied , which completes the proof.

■■■

<u>Remarks</u>

(1) is worth mentioning that the proposed hierarchical control structure is different from that developed in [85-87] due to the following reasons :

(i) The second term of the present control (the coordinator gain) is nonlinear one and depends upon the initial conditions.

(ii) We are still working on subsystem level except for $\sigma(\underline{y})$ which needs some information about the other subsystems. However, in the light of (6.35) , the calculation of $\sigma(\underline{y})$ is done for each subsystem separately.

(iii) The exchange of information , in the present control implementation , between the coordinator and the subsystems is minimal and hence the control is easy to be realized especially for large - scale , geographically distributed systems.

(2) Again , in the case when $\underline{w}_i = \underline{0}$ and $\epsilon = 0$, it follows from (6.38) , with the aid of (6.37) and (6.39) , that $\bar{\eta} = 0$ and hence $\Omega_c(\bar{\eta})$ will represent the whole space.

6.6 Stability of The System Under Structural Perturbations

In general , the implementation of hierarchical control necessitates the use of many communication links in order to facilitate transmitting pertinent information and corrective signals. However , in many situations physical perturbations took place which tend to modify the data transmission network and cause information losses. This may affect the stability of the global system and frequently undergoes unstability [85-87]. In the sequel , we will present three theorems to investigate the stability of the overall uncertain system when subjected to worst case structural perturbations.

6.6.1 Structural Perturbation Between The Coordinator and Subsystems

We assume the worst case ; that is the corrective signals (nonlinear control part) , feeding the subsystems from the coordinator , are completely absent , i.e. $\underline{g}(\underline{y},t) = \underline{0}$. In this case , the sufficient conditions for the stabilizability of the global system is given by the following theorem :

Theorem 6.6.1

In the absence of the interconnection links between the two levels , the system will be practically stable if

(i) the test matrix $L = [l_{ij}]$ given by (6.16) is positive definite matrix

(ii) $\lambda_m(L) > 2\,\tilde{b}\,\|B^T P\|$ (6.45)

where $\tilde{b}$ is given by (6.37b).

Moreover , the resulting closed-loop state trajectories are bounded in a domain $\Omega^c(\eta_1)$ where

$$\eta_1 = 2\,\tilde{a}\,\|B^T P\| \,/\, [\,\lambda_m(L) - 2\,\tilde{b}\,\|B^T P\|\,] \qquad (6.46)$$

where a is given by (6.37a).

Proof

By putting $\underline{g}(\underline{y}(t),t) = \underline{0}$ in (6.32) and following the analysis pursued from (6.40) to (6.44). ■■■

6.2 Structural Perturbation Between The Subsystems

In this case , it will be assumed that the disturbances take place in the interconnection links between the subsystems while the overall information structure is maintained without any changes. Also , we suppose the worst case design when the interactions among the subsystems are completely ignored ; that is $\underline{Z}_i = \underline{h}_i = \underline{0}$; i=1,2,......,N.

Theorem 6.6.2

If the decomposed subsystems are completely decoupled , then the global system is practically stable in a domain of attraction $\Omega^c(\eta_2)$ with radius η_2 given by

$$\eta_2 = [\,\bar{\mu}_1 + \sqrt{\bar{\mu}_1^2 + 4\;\bar{\mu}_0\;\lambda_m(Q)}\,] \,/\, [2\,\lambda_m(Q)] \qquad (6.47)$$

where $\bar{\mu}_0$ and $\bar{\mu}_1$ are given by (6.39).

Proof

It can be easily obtained by putting $M = H(t,\underline{r},\underline{x}) = 0$ in (6.40) and following analysis similar to (6.41) - (6.44).

6.6.3 Structural Perturbation Between The Coordinator and Subsystems and Cutting The Links Between The Subsystems

This situation considers the previous two cases simultaneously ; that is $\underline{g}(\underline{y},t) = \underline{Z}_i = \underline{h}_i = \underline{0}$; $i=1,2,\ldots,N$.

Theorem 6.6.3

When all the links between the coordinator and subsystems and all interactions among the subsystems are cut , then the overall system is practically stable in a domain of attraction $\Omega_c(\eta_3)$ if the following condition holds :

$$\lambda_m(Q) > 2\,\tilde{b}\,\|B^T P\| \tag{6.48}$$

where

$$\eta_3 = 2\,\tilde{a}\,\|B^T P\| \,/\, [\,\lambda_m(Q) - 2\,\tilde{b}\,\|B^T P\|\,] \tag{6.49}$$

and $\tilde{a}$,$\tilde{b}$ are given by (6.37).

Proof

The theorem can be proved by setting $\underline{g}(\underline{y}(t),t) = \underline{0}$ in (6.32) and $M = H(t,\underline{r},\underline{x}) = 0$ in (6.40) and applying the Lyapunov analysis of (6.41) through (6.44). ▪▪▪

Remark 6.6.1

Theorems (6.6.1) and (6.6.3) establish the efficacy of linear control in stabilizing interconnected subsystems and decomposed subsystems respectively.

Remark 6.6.2

Again, if $\underline{w}_i = \underline{0}$ and $\epsilon = 0$, it follows from (6.41), (6.42) and (6.47) that $\eta_1 = \eta_2 = \eta_3 = 0$ and hence the domain of stability in the three cases will represent the whole space.

6.7 Illustrative Example

The six plate gas absorber system [92,94] is considered. It is assumed that the system is a combination of two subsystems as in Figure 6.2, each is composed of 3-plates (i.e. each subsystem is a 3-plate gas absorber). Furthermore, x_3 and x_4 represent the interconnection between the two subsystems. The system dynamics are given by [92] :

$$\dot{\underline{x}}(t) = A\,\underline{x}(t) + B\,\underline{u}(t) \tag{6.50}$$

where $\underline{x}(t) \in R^6$, $\underline{u}(t) \in R^2$, with

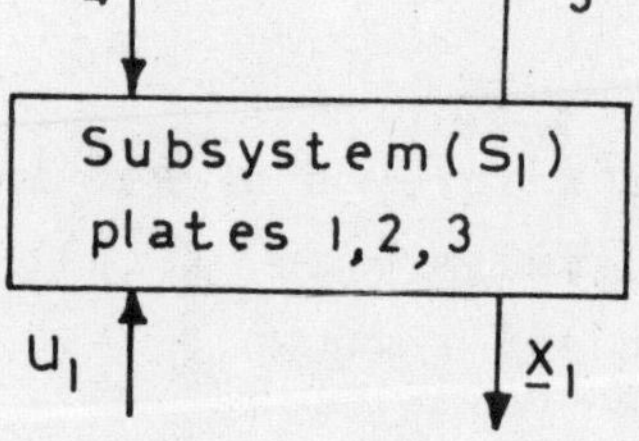

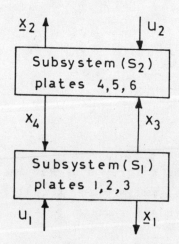

Figure 6.2: Six-plate Gas Absorber Representation.

$$A = \begin{bmatrix} -d_2(1+d_1) & d_2 & 0.0 & & & 0.0 \\ d_1 d_2 & -d_2(1+d_1) & d_2 & 0.0 & & \\ 0.0 & d_1 d_2 & -d_2(1+d_1) & d_2 & 0.0 & \\ & & & & & \\ & & & & & 0.0 \\ 0.0 & & & 0.0 & d_1 d_2 & -d_2(1+d_1) \end{bmatrix}$$

$$B^T = \begin{bmatrix} d_1 d_2 & 0.0 & & 0.0 \\ 0.0 & & & d_2 / a_o \end{bmatrix}$$

The choice of parameter values $d_1 = 0.849$, $d_2 = 0.634$ and a_0 = constant = 0.72 yields the matrices of the system (6.45) of the form [92,94] :

$$A = \left[\begin{array}{c|c} A_1 & A_{12} \\ \hline A_{21} & A_2 \end{array}\right]$$

$$A = \left[\begin{array}{ccc|ccc} -1.173 & 0.634 & 0.0 & 0.0 & 0.0 & 0.0 \\ 0.538 & -1.173 & 0.634 & 0.0 & 0.0 & 0.0 \\ 0.0 & 0.538 & -1.173 & 0.634 & 0.0 & 0.0 \\ \hline 0.0 & 0.0 & 0.538 & -1.173 & 0.634 & 0.0 \\ 0.0 & 0.0 & 0.0 & 0.538 & -1.173 & 0.634 \\ 0.0 & 0.0 & 0.0 & 0.0 & 0.538 & -1.173 \end{array}\right]$$

$$B = \left[\begin{array}{c|c} B_1 & 0 \\ \hline 0 & B_2 \end{array}\right] = \left[\begin{array}{c|c} 0.538 & 0.0 \\ 0.0 & 0.0 \\ 0.0 & 0.0 \\ \hline 0.0 & 0.0 \\ 0.0 & 0.0 \\ 0.0 & 0.88 \end{array}\right]$$

We consider that the parameters d_1 and d_2 undergo ± 25% variation about their nominal values. We thus have :

$$\Delta A(\underline{r}) = \left[\begin{array}{c|c} \Delta A_1(\underline{r}) & 0 \\ \hline 0 & \Delta A_2(\underline{r}) \end{array}\right]$$

$$\Delta B(\underline{s}) = \left[\begin{array}{c|c} \Delta B_1(s_1) & 0 \\ \hline 0 & \Delta B_2(s_2) \end{array}\right]$$

$$H(t,\underline{r},\underline{x}) = \left[\begin{array}{c|c} 0 & \Delta A_{12}(\underline{r}) \\ \hline \Delta A_{21}(\underline{r}) & 0 \end{array}\right]$$

where

$$\Delta A_1(\underline{r}) = \Delta A_2(\underline{r}) = \begin{bmatrix} r_1 & r_2 & 0.0 \\ r_3 & r_1 & r_2 \\ 0.0 & r_3 & r_1 \end{bmatrix}$$

$$\Delta B_1(s_1) = \begin{bmatrix} s_1 \\ 0.0 \\ 0.0 \end{bmatrix} , \quad \Delta B_2(s_2) = \begin{bmatrix} 0.0 \\ 0.0 \\ s_2 \end{bmatrix}$$

$$\Delta A_{12}(\underline{r}) = \begin{bmatrix} 0.0 & 0.0 & 0.0 \\ 0.0 & 0.0 & 0.0 \\ r_2 & 0.0 & 0.0 \end{bmatrix} , \quad \Delta A_{21}(\underline{r}) = \begin{bmatrix} 0.0 & 0.0 & r_3 \\ 0.0 & 0.0 & 0.0 \\ 0.0 & 0.0 & 0.0 \end{bmatrix}$$

and the compact bounding sets $\mathcal{R}_i$ and $\mathcal{S}_i$, (i=1,2) are given by :

$$\mathcal{R}_1 = \mathcal{R}_2 = \{ \underline{r} \in R^3 : \begin{array}{l} -0.46 \le r_1 \le 0.39 , \\ -0.158 \le r_2 \le 0.158 , \\ -0.235 \le r_3 \le 0.303 \} \end{array}$$

$$\mathcal{S}_1 = \{ s_1 : -0.235 \le s_1 \le 0.303 \}$$

$$\mathcal{S}_2 = \{ s_2 : -0.219 \le s_2 \le 0.221 \}$$

It is worth mentioning that the matching conditions are met for the two subsystems with :

$$D_1(\underline{r}) = D_2(\underline{r}) = 1.859 \, [r_1 \quad r_2 \quad 0.0]$$

$$E_1(s_1) = 1.859 \, s_1 \quad , \quad E_2(s_2) = 1.136 \, s_2$$

and hence $\rho_{r1} = \rho_{r2} = 0.904$, $\rho_{s1} = 0.563$, $\rho_{s2} = 0.251$
Note that $\rho_{s1} < 1$, $\rho_{s2} < 1$ satisfying Assumption 5(iii).

In order to evaluate G_1 and G_2 , we apply the decentralized control theory with unity weighting matrices and degree of stability $\alpha = 1.6$ [93] and hence the results are :

$$G_1 = [-6.6712 \quad -16.3014 \quad -17.1622] \tag{6.51a}$$

$$G_2 = [-9.24 \quad -9.944 \quad -4.4264] \tag{6.51b}$$

Thus

$$\tilde{A}_1 \triangleq A_1 + B_1 G_1 = \begin{bmatrix} -4.7621 & -8.1362 & -9.2333 \\ 0.538 & -1.173 & 0.634 \\ 0.0 & 0.538 & -1.173 \end{bmatrix}$$

$$\tilde{A}_2 \triangleq A_2 + B_2 G_2 = \begin{bmatrix} -1.173 & 0.634 & 0.0 \\ 0.538 & -1.173 & 0.634 \\ -8.1312 & -8.2127 & -5.0684 \end{bmatrix}$$

With $Q_1 = Q_2 = I_3$, the solution of (6.9), i=1,2 are :

$$P_1 = \begin{bmatrix} 0.1007 & -0.0381 & -0.0663 \\ -0.0381 & 1.1686 & 1.0421 \\ -0.0663 & 1.0421 & 1.511 \end{bmatrix}$$

$$P_2 = \begin{bmatrix} 1.1101 & 0.788 & -0.0465 \\ 0.788 & 1.0312 & -0.0256 \\ -0.0465 & -0.0256 & 0.0955 \end{bmatrix}$$

Proceeding further , we compute $\sigma_1(\underline{x}_1)$ and $\sigma_2(\underline{x}_2)$ as :

$\sigma_1(\underline{x}_1) = 33.75 \; \| \underline{x}_1 \|$, $\sigma_2(\underline{x}_2) = 5.991 \; \| \underline{x}_2 \|$

where $\underline{x}_1 = [\, x_1 \,,\, x_2 \,,\, x_3 \,]^T$, $\underline{x}_2 = [\, x_4 \,,\, x_5 \,,\, x_6 \,]^T$

With $\epsilon_1 = \epsilon_2 = 0.5$, the nonlinear term is given by :

$\underline{g}(\underline{x}) = [\, g_1(\underline{x}_1) \;,\; g_2(\underline{x}_2) \,]^T$, where for i=1,2

$g_i(x_i) =$

$$\begin{cases} -\sigma(\underline{x}_i)\, B_i^T P_i \underline{x}_i \,/\, \| B_i^T P_i \underline{x}_i \| & \text{for } \| B_i^T P_i \underline{x}_i \| > 0.5 \\ -\sigma(\underline{x}_i)\, B_i^T P_i \underline{x}_i \,/\, 0.5 & \text{for } \| B_i^T P_i \underline{x}_i \| \le 0.5 \end{cases} \qquad (6.52)$$

This system was simulated in order to study its dynamic behaviour under normal operating conditions as well as under various structural perturbations either between the subsystems and/or in the information network.

6.7.1 The Decentralized Solution

The problem at hand is solved in order to generate the decentralized control structure shown in Figure 6.3.

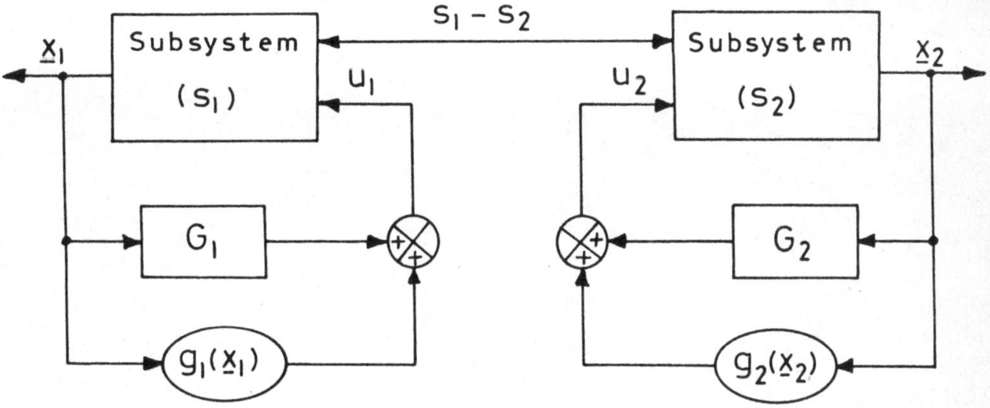

Figure 6.3: The Decentralized Control Structure.

Where G_1 and G_2 are given by (6.51) whereas $g_1(\underline{x}_1)$ and $g_2(\underline{x}_2)$ are given by (6.52).

For the purpose of comparison , our system is run for the following three cases :

(a) $r_1 = 0.2$, $r_2 = 0.1$, $r_3 = 0.25$, $s_1 = 0.28$, $s_2 = 0.2$

(b) $r_1 = -0.46$, $r_2 = -0.158$, $r_3 = s_1 = -0.235$, $s_2 = -0.219$

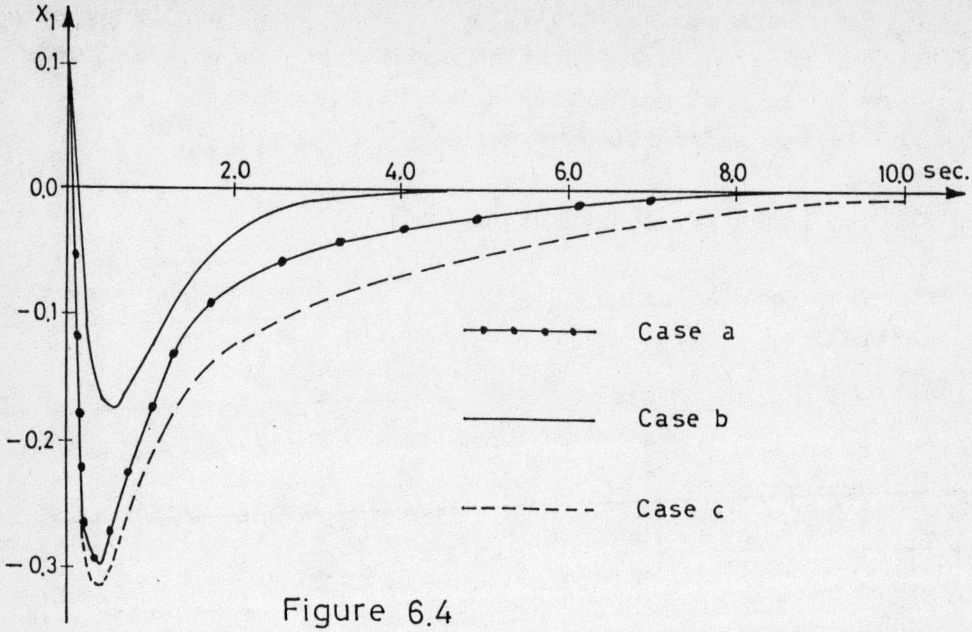

Figure 6.4

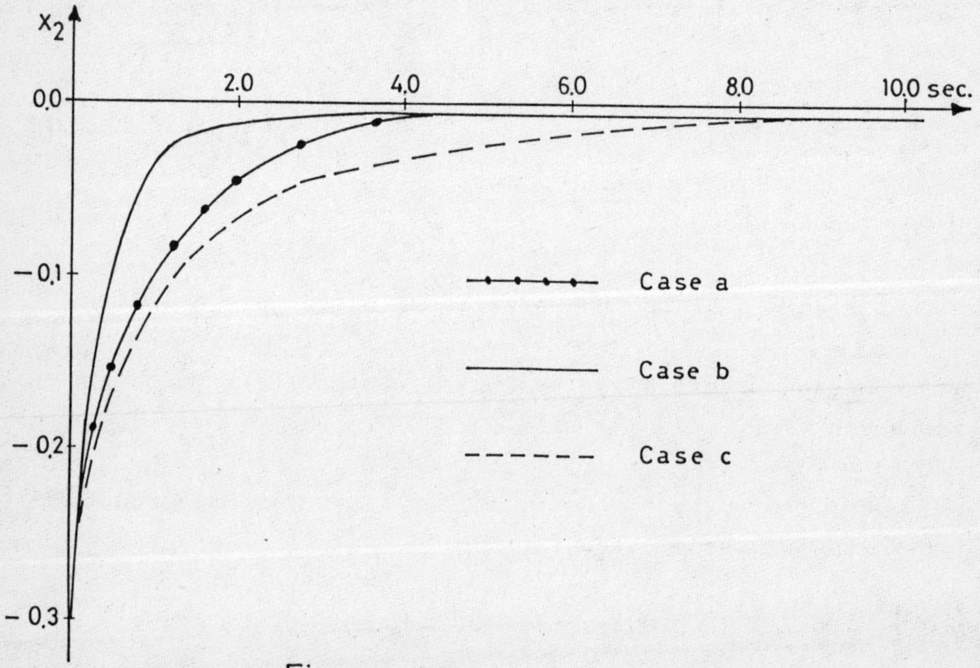

Figure 6.5

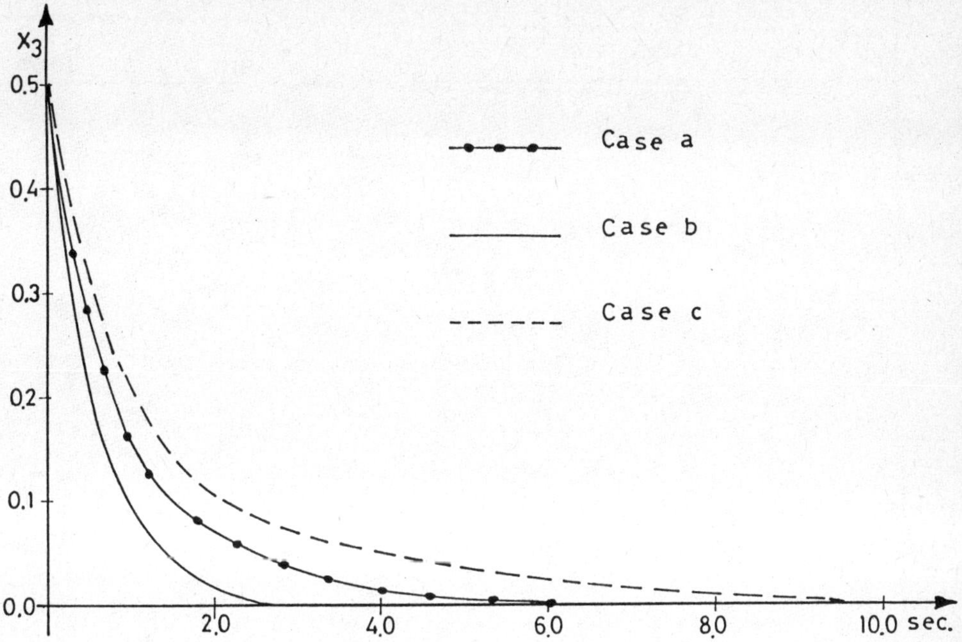

Figure 6.6

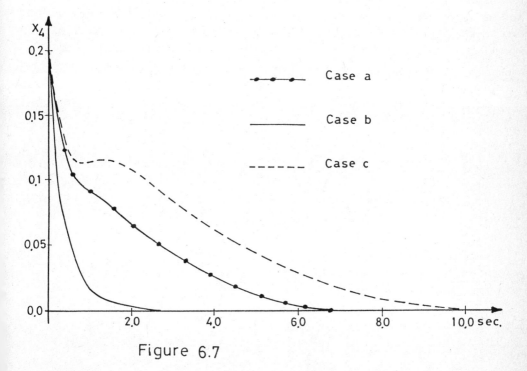

Figure 6.7

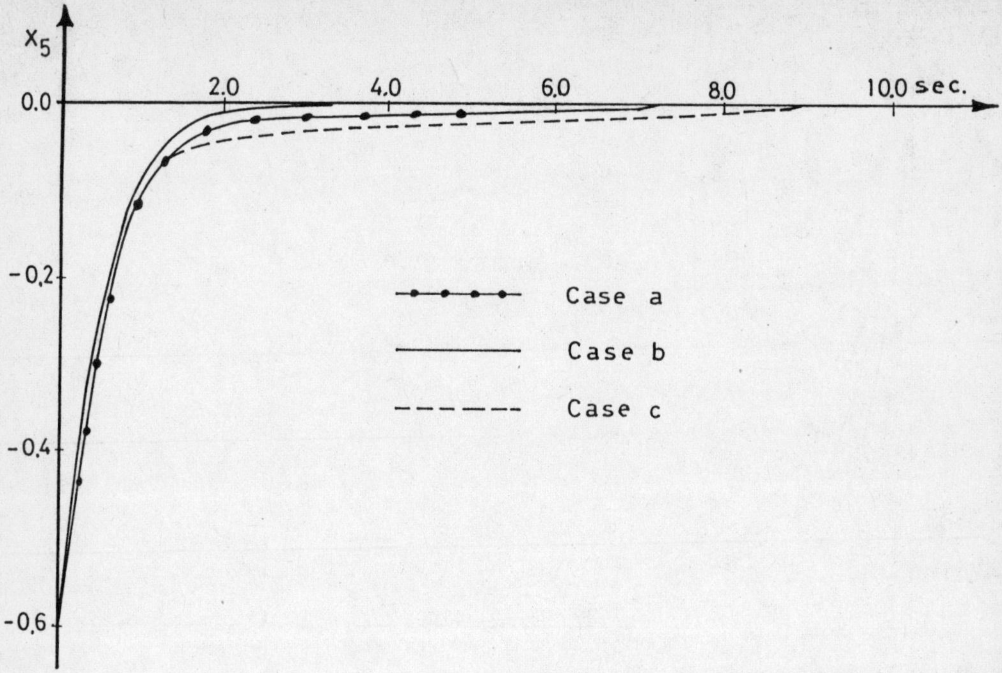

Figure 6.8

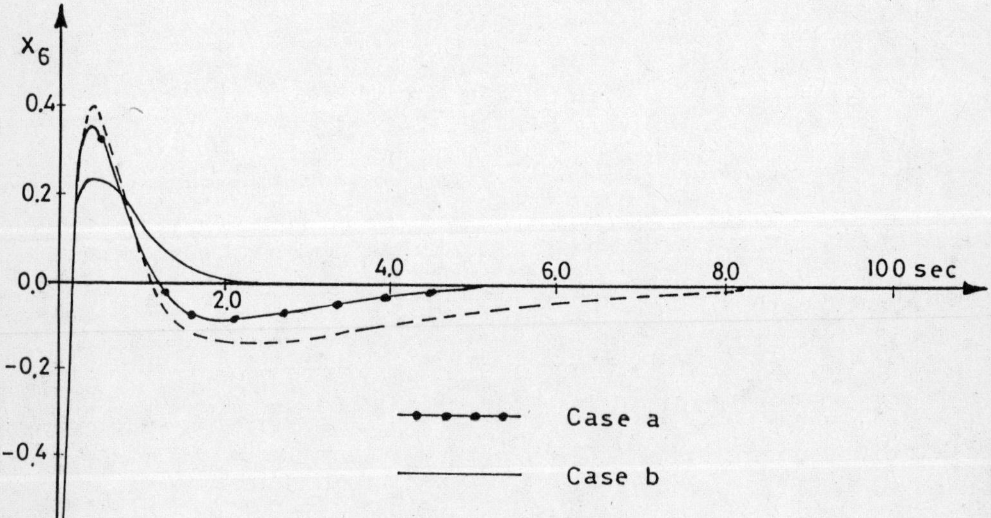

Figure 6.9

(c) r_1 = 0.39 , r_2 = 0.158 r_3 = s_1 = 0.303 , s_2 = 0.221

and the three cases are plotted on the same graphs shown in Figures (6.4-6.9).

6.7.2 The Hierarchical Solution for The Interconnected System

The problem is solved now in order to generate the following hierarchical structure shown in Figure 6.10.

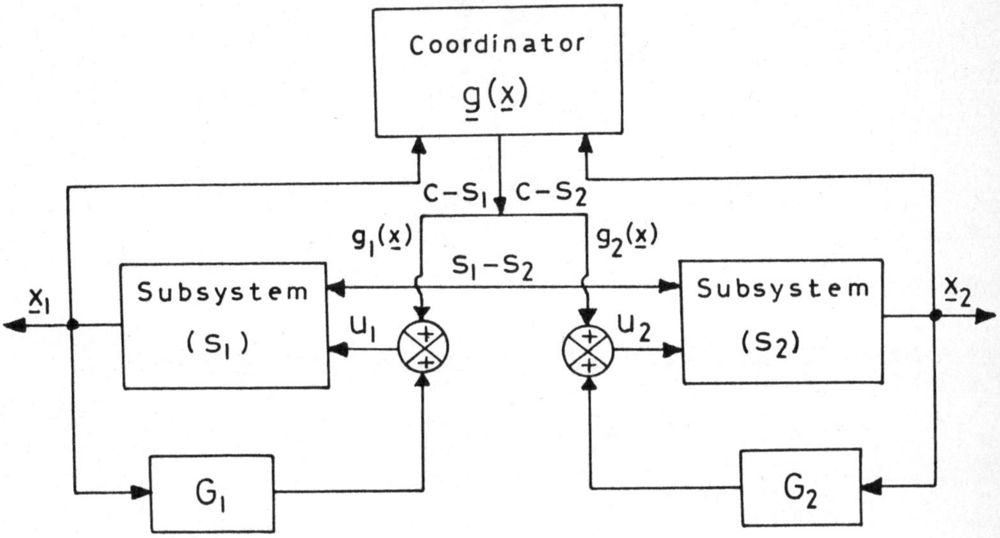

Figure 6.10: The Hierarchical Control Structure.

Where G_1, G_2 are given by (6.46) and $g_1(\underline{x})$, $g_2(\underline{x})$ are given by (6.47) with

$$\sigma_1(\underline{x}) = \sigma_2(\underline{x}) = \sigma(\underline{x}) = 33.75 \; \| \underline{x} \| \; ;$$

where $\underline{x} = (\underline{x}_1^T \; , \; \underline{x}_2^T)^T$

The resulting closed-loop state trajectories , with the values of r and s given in (6.7.1b), are shown in Figures (6.11-6.16).

We performed the following structural perturbations between the coordinator and subsystems :

(a) link (C - S_1) is cut , i.e. $g_1(\underline{x}) = 0$.

(b) link (C - S_2) is cut , i.e. $g_2(\underline{x}) = 0$.

(c) the two links (C - S_1) and (C - S_2) are removed.

For the purpose of comparison , the above three cases are plotted on the same graphs depicted in Figures (6.11-6.16).

6.7.3 The Hierarchical Solution for The Decoupled Subsystems

In this case , the hierarchical control structure is as shown in Figure 6.10 except the interactions (S_1 - S_2) being cut. Furthermore, the same simulations, of subsection 6.7.2 , are run again and plotted in Figures 6.17 - 6.22.

From these simulations , one can conclude that the resulted closed-loop uncertain system in each case is stable.

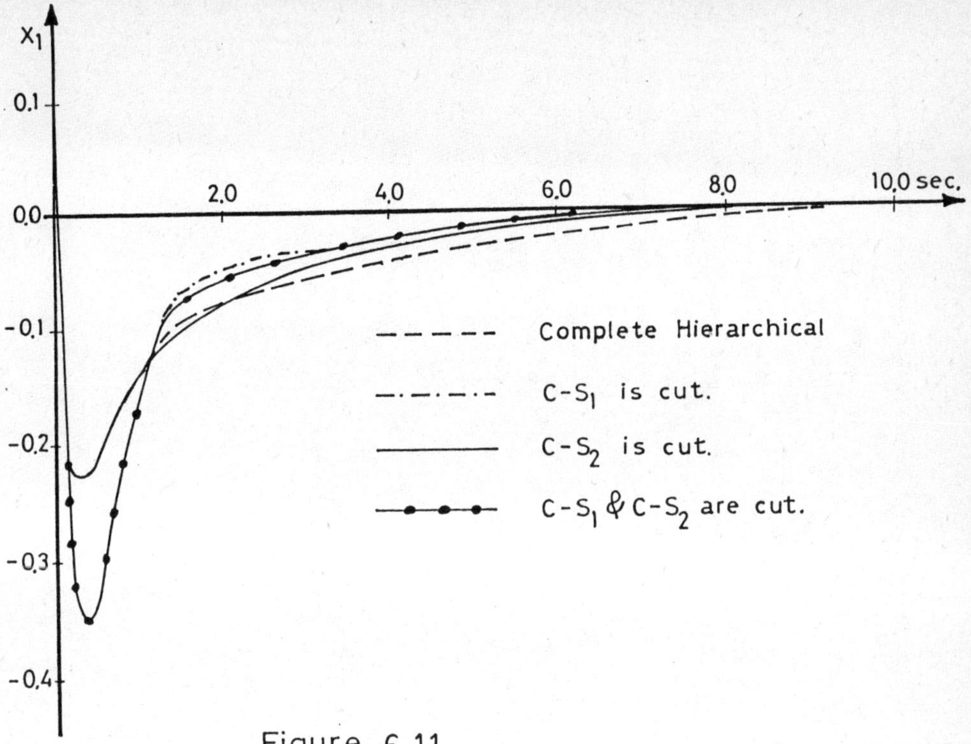

Figure 6.11

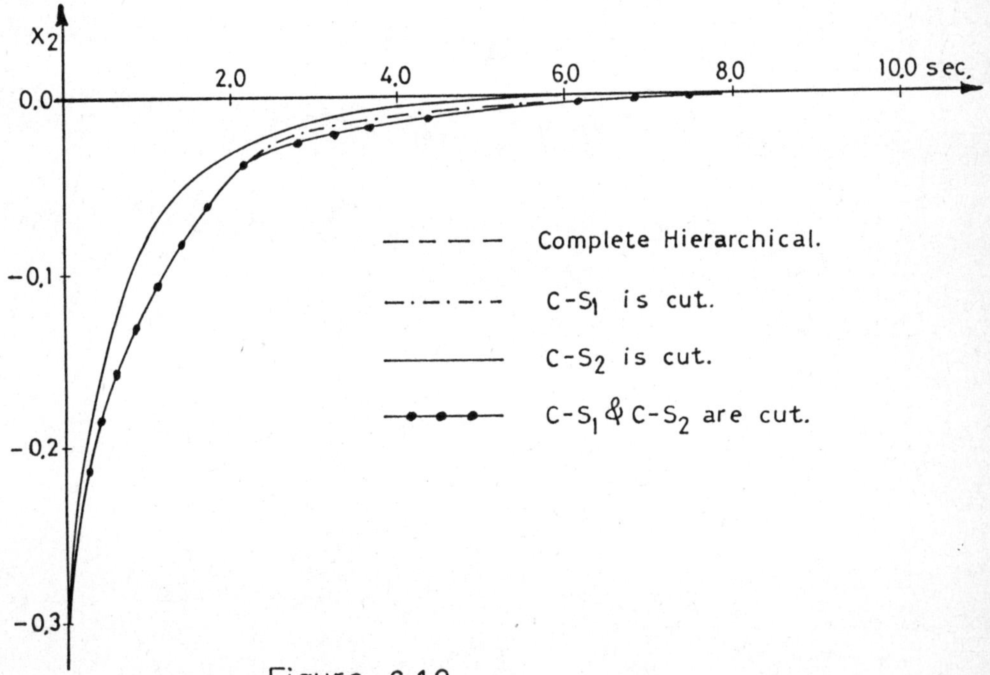

Figure 6.12

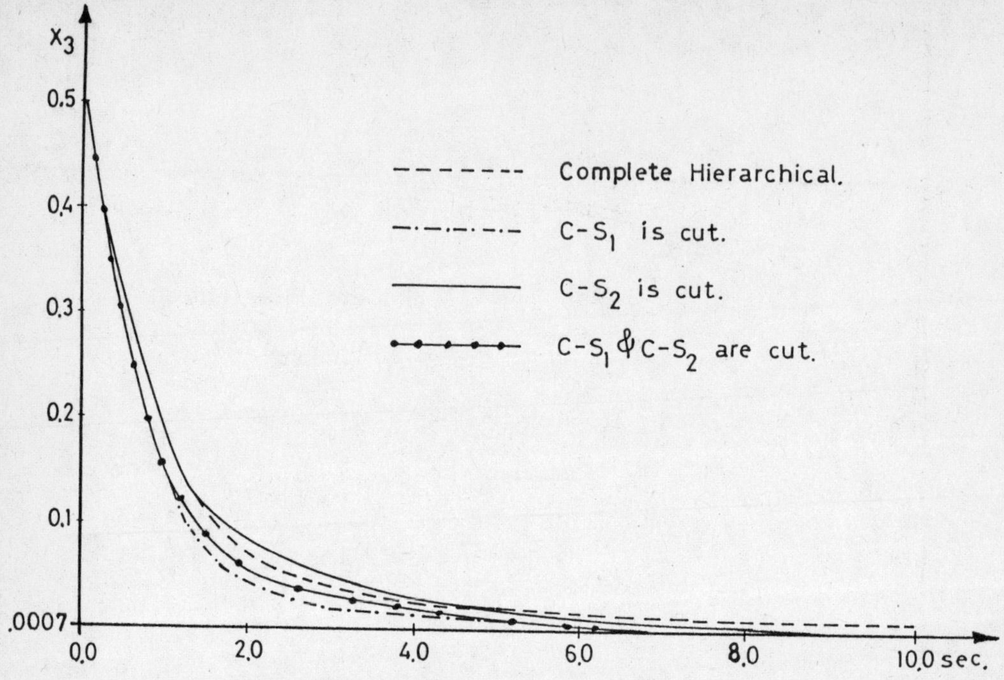

Figure 6.13

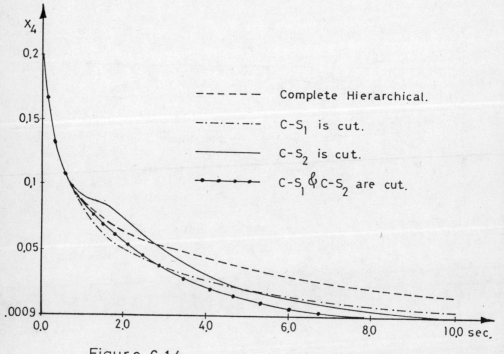

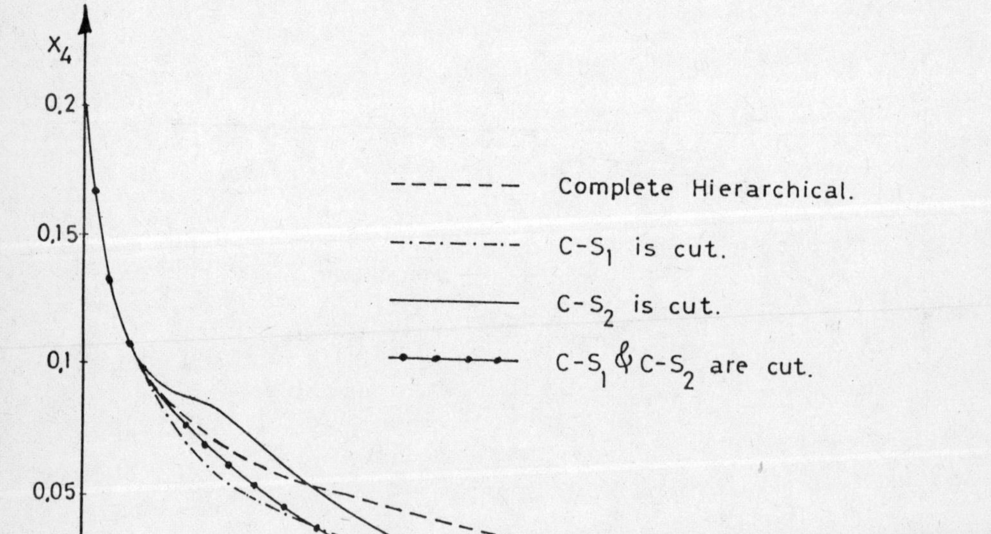

Figure 6.14

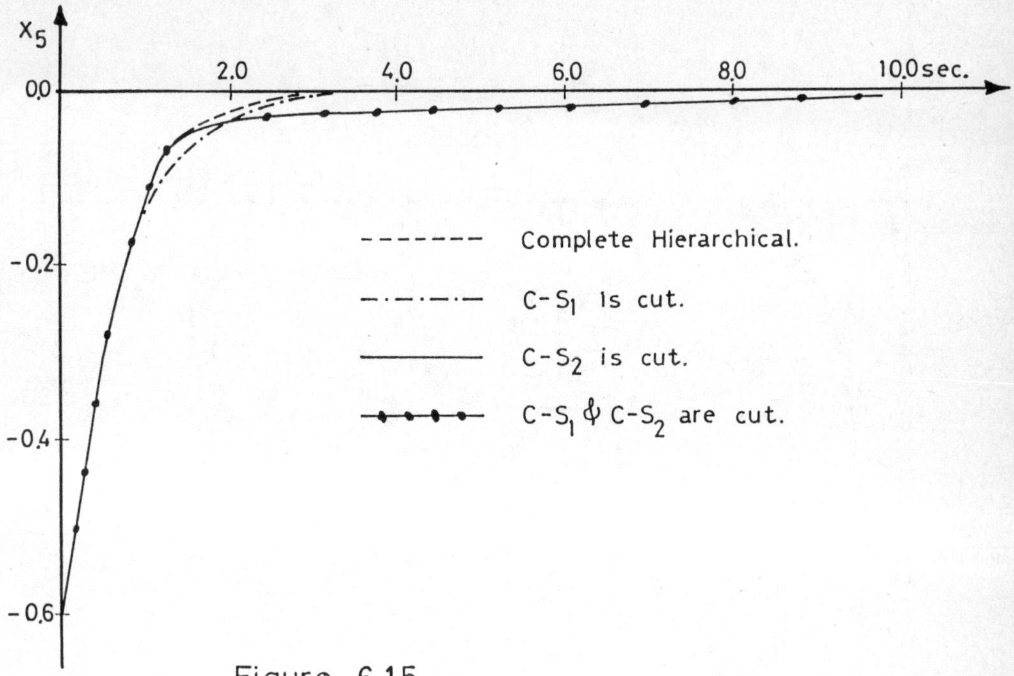

Figure 6.15

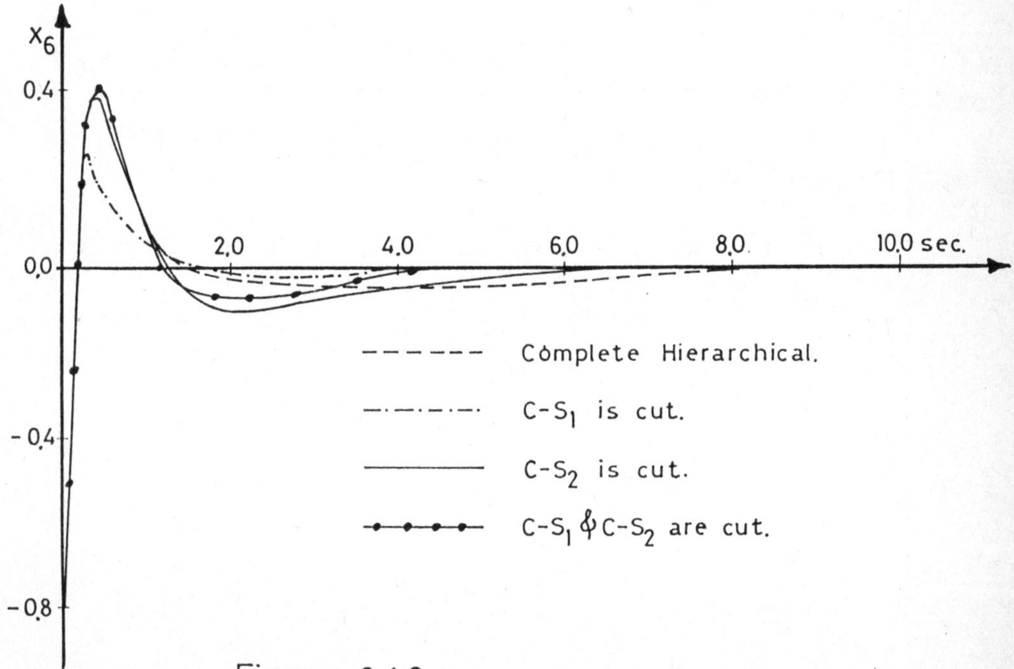

Figure 6.16

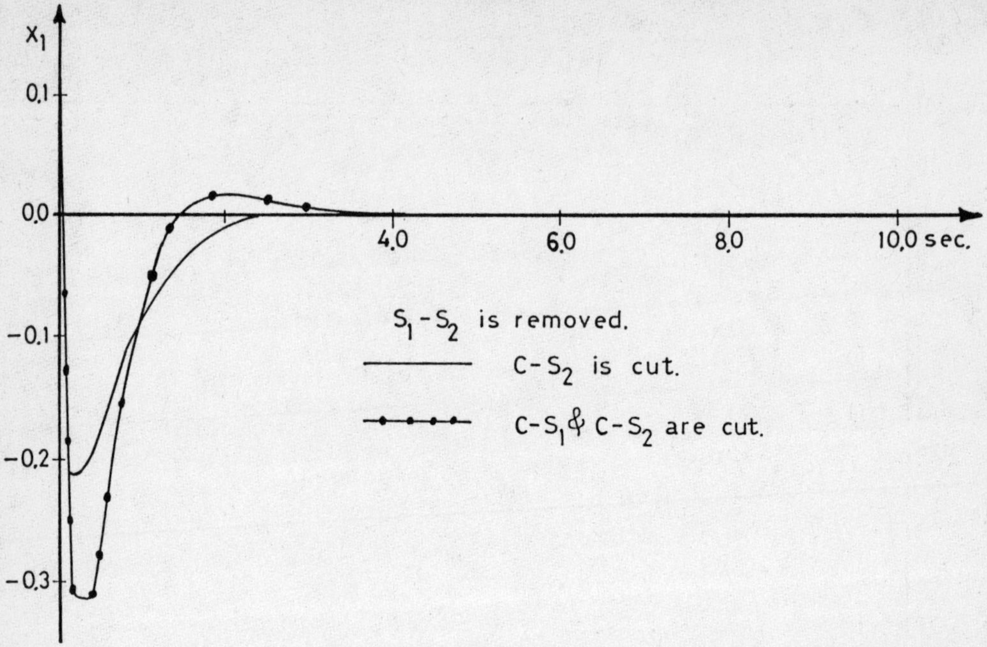

Figure 6.17

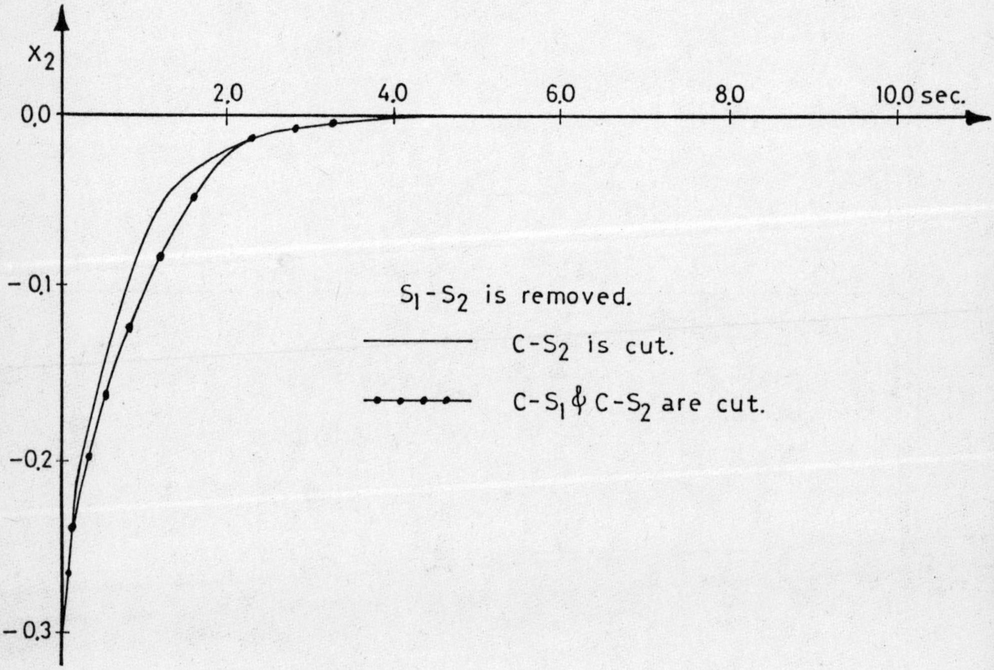

Figure 6.18

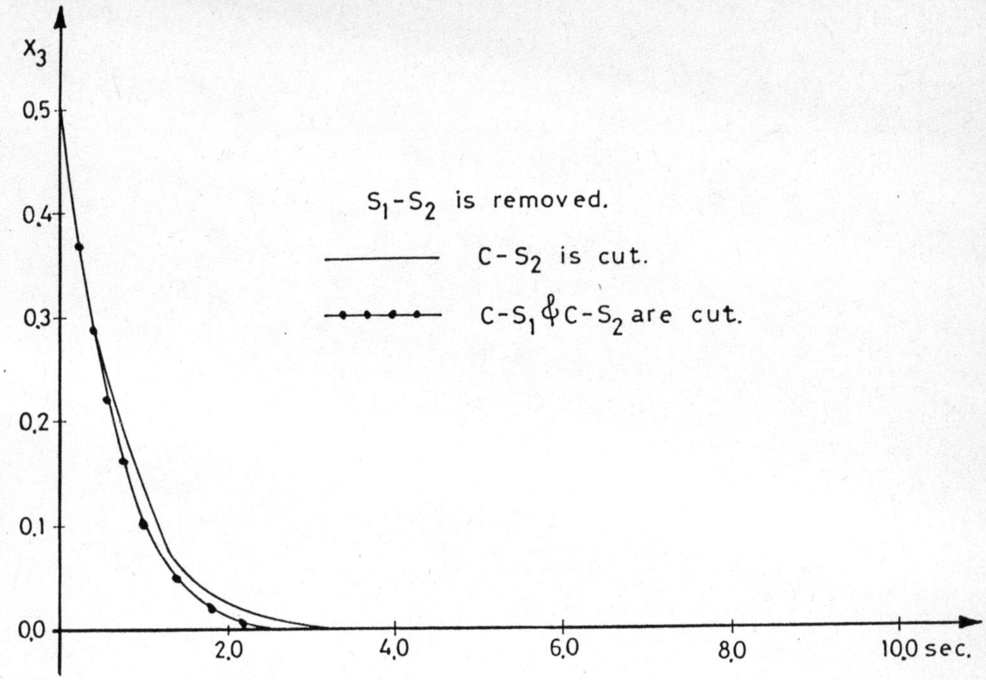

Figure 6.19

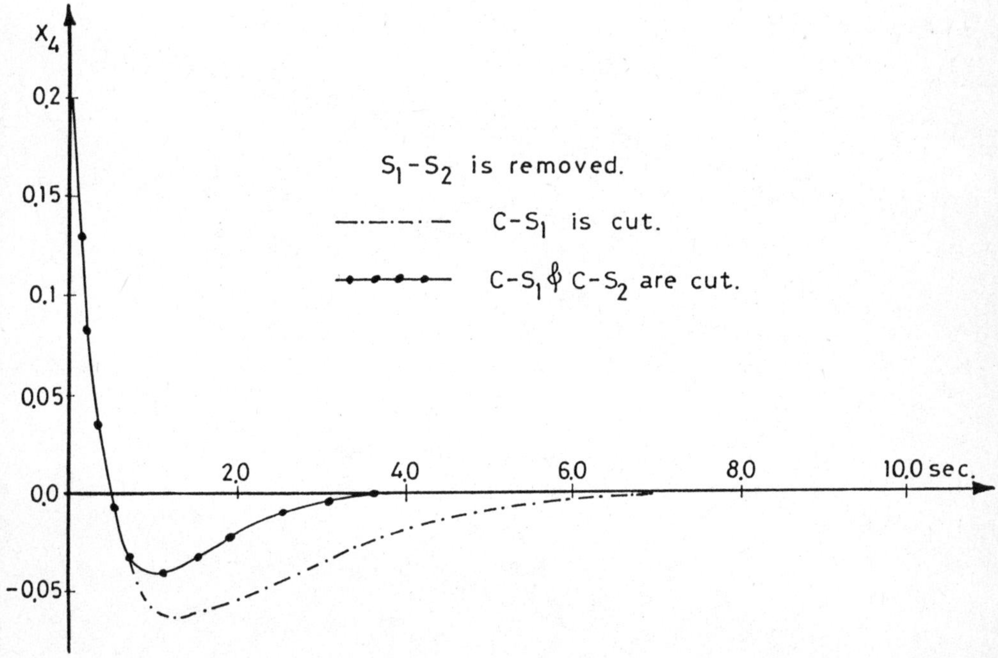

Figure 6.20

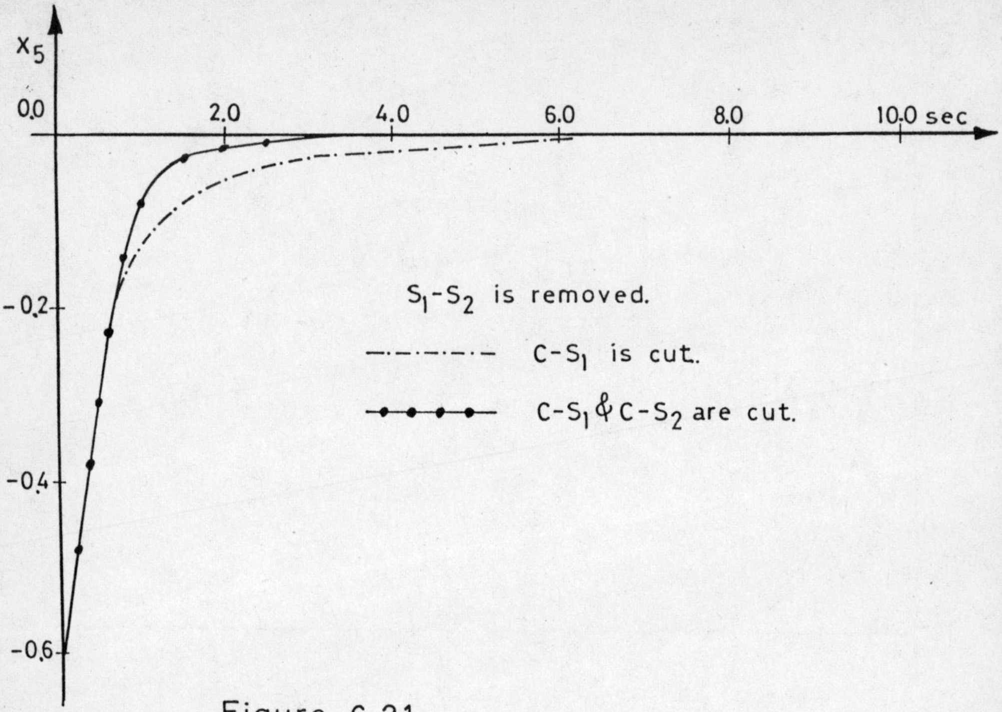

Figure 6.21

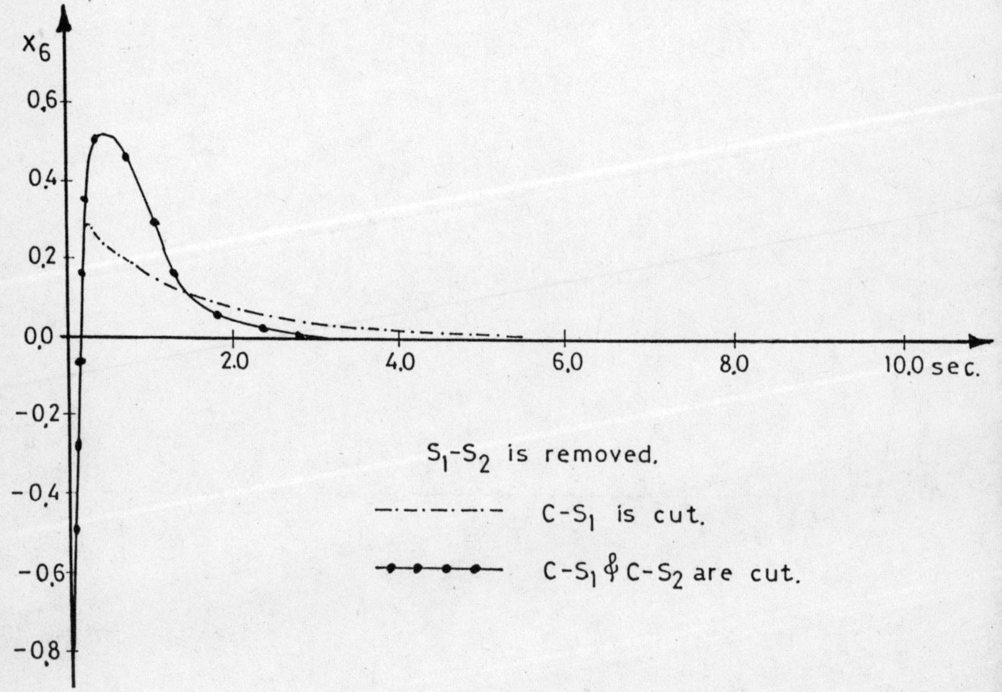

Figure 6.22

6.8 Conclusions

In this Chapter, new decentralized and hierarchical control schemes for large-scale interconnected , uncertain dynamical systems have been presented. From the derived analysis and simulation results , the developed approaches have the following main features :

(i) The control structures are very easy to compute since all calculations are realized at the subsystem level. Moreover , they can be implemented using available hardware assembly.

(ii) The decentralized control approach , under certain hypothesis , results in a stable overall uncertain system despite the complete absence of liaison between the local control stations.

(iii) With the hierarchical control scheme , one can guarantee the boundedness behaviour of the global system even in the presence of structural perturbations that may occur either in the interconnections between the subsystems and/or in the data transmission network between the two levels of the structure.

(iv) The obtained results prevent or at least reduce the transfer of pertinent information between the subsystems. This is a great advantage specially for large-scale systems characterized by their geographically scattered nature , such as power systems , telephone networks ,....,etc.

CHAPTER 7

INTERCONNECTED SYSTEMS : DISCRETE CASE

7.1 Introduction

One of the principle applications of feedback in control systems is to reduce the deviations that may arise in desired system response due to "plant parameter uncertainty" and "uncertainty of external inputs". In Chapter 6, we discussed how to design feedback control schemes for large-scale, continuous systems with unknown but bounded parameters. It has been shown that with the satisfaction of certain sufficient conditions and/or the validity of the uncertainty matching structure, the developed techniques provide uniform ultimate boundedness of the system at hand.

However, the rapid growth in computing capabilities and the improved technology of microprocessors has attracted system analysts and modellers to utilize digital computers extensively in solving their problems. This is the case with many industrial processes where digital devices are often used. In such industrial applications, we have batch information processing in contrast to the continuous information processing which was required when traditional analog equipment was used [2].

So, we begin in the next section to formulate the problem of stabilizing a class of discrete-time interconnected uncertain dynamical systems. Then, a complete decentralized scheme is developed in sections 7.3 and 7.4 in order to stabilize the system at hand. Development of an efficient hierarchical controller in order to regulate the overall system is included in section 7.5. Hence, the boundedness of the integrated system under various structural perturbations has been tested in section 7.6. In all situations, the stability boundedness behaviour has been clarified. In section 7.7, a mechanical manipulator control system example is employed as a vehicle to illustrate the developed approaches. Uncertainty of the system description is included as a feature of the example. Finally, we conclude in section 7.8 with a brief discussion of the relevance of the developed results.

7.2 Problem Statement

Consider a class of uncertain dynamical systems shown in Figure 7.1 and described as an interconnection of N subsystems ,i.e.

$$\underline{x}_i(k+1) = [A_i + \Delta A_i(\underline{r}_i(k))]\,\underline{x}_i(k) + [B_i + \Delta B_i(\underline{s}_i(k))]\,\underline{u}_i(k)$$

$$+ C_i\,\underline{v}_i(k) + \underline{\zeta}_i + \underline{h}_i \quad ; \; i=1,2,\ldots,N \tag{7.1a}$$

$$\underline{\zeta}_i = \sum_{\substack{j=1 \\ j\neq i}}^{N} A_{ij}\,\underline{x}_j \tag{7.1b}$$

$$\underline{h}_i = \sum_{\substack{j=1 \\ j\neq i}}^{N} H_{ij}(k, \underline{r}_i(k), \underline{x}_j(k)) \tag{7.1c}$$

$$\underline{x}_i(k_0) = \underline{x}_{i0} \tag{7.1d}$$

$$\underline{y}_i(k) = \underline{x}_i(k) + \underline{w}_i(k) \tag{7.2}$$

where for the ith subsystem : $\underline{x}_i \in R^{n_i}$ is the state , $\underline{u}_i \in R^{m_i}$ is the control , $\underline{v}_i \in R^{l_i}$ is bounded disturbance , $\underline{y}_i \in R^{n_i}$ is the measured state , $\underline{w}_i \in R^{n_i}$ is the measurement error ,

$\underline{z}_i \in R^{n_i}$ is the interconnection with the other subsystems and the constant matrices A_i , B_i , C_i , A_{ij} are prescribed with appropriate dimensions. The uncertainty parameters are $\underline{r}_i \in \mathcal{R}_i \subset R^{p_i}$, $\underline{s}_i \in \mathcal{S}_i \subset R^{q_i}$, $\underline{v}_i \in \mathcal{V}_i \subset R^{l_i}$ and $\underline{w}_i \in \mathcal{W}_i \subset R^{n_i}$. The subsystem matrix uncertainty $\Delta A_i(\underline{r}_i)$ and the input component matrix uncertainty $\Delta B_i(\underline{s}_i)$ depend on parameters $\underline{r}_i$ and $\underline{s}_i$ respectively. The term $C_i\ \underline{v}_i(k)$ accounts for external input uncertainty. Moreover , $\underline{h}_i \in R^{n_i}$ contains the uncertainties in the interconnections between the ith subsystem and the other subsystems.

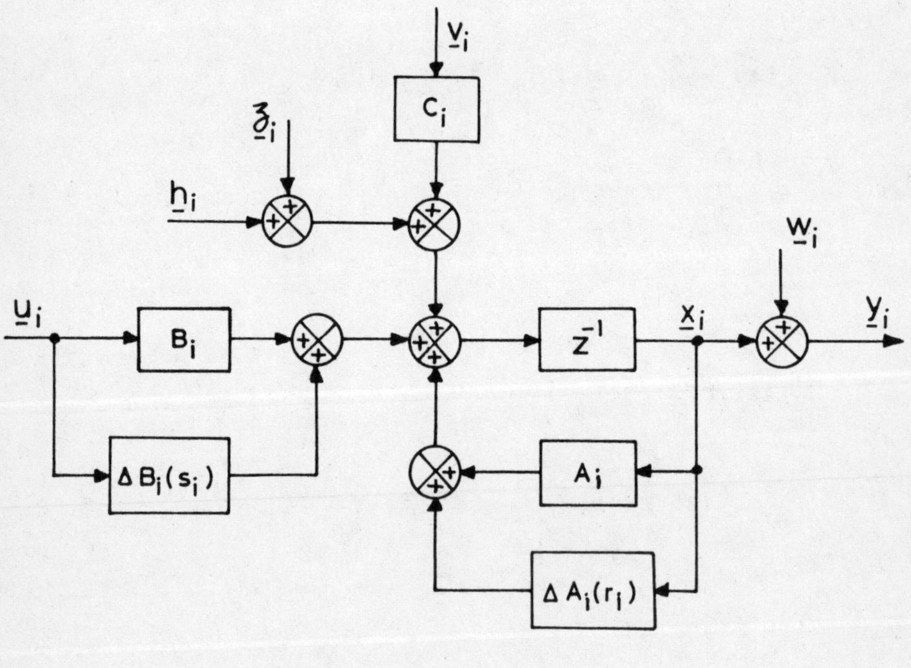

Figure 7.1: Additive-type Uncertain Discrete Subsystem.

Throughout the remainder of this Chapter, the following assumptions are taken for each subsystem as standard [85-92]:

(A1) The entries of $\Delta A_i(.)$ and $\Delta B_i(.)$ are continuous on R^p and R^q respectively.

(A2) The bounding sets $\mathcal{R}_i$, $\mathcal{S}_i$, $\mathcal{V}_i$ and $\mathcal{W}_i$ are compact sets in the indicated finite-dimensional Euclidean spaces.

(A3) The uncertainties $\underline{r}_i(.)$: $R \longrightarrow \mathcal{R}_i$, $\underline{s}_i(.)$: $R \longrightarrow \mathcal{S}_i$, $\underline{v}_i(.)$: $R \longrightarrow \mathcal{V}_i$ and $\underline{w}_i(.)$: $R \longrightarrow \mathcal{W}_i$ are assumed to be Lebesgue measurable.

(A4) The pair (A_i, B_i) is stabilizable ; that is there exists, a constant $(m_i \times n_i)$ matrix G_i, such that the eigenvalues $\lambda(\tilde{A}_i) \triangleq (A_i + B_i G_i)$ have moduli strictly less than unity.

(A5) The following matching conditions are met :

(i) There exist matrix functions (of appropriate dimensions) $D_i(.)$ and $E_i(.)$ whose entries are continuous on R^p and R^q respectively such that :

$$\Delta A_i(\underline{r}_i) = B_i \; D_i(\underline{r}_i) \tag{7.3a}$$

$$\Delta B_i(\underline{s}_i) = B_i \; E_i(\underline{s}_i) \tag{7.3b}$$

(ii) There exists a constant matrix function F_i such that

$$C_i = B_i \; F_i \tag{7.3c}$$

(iii) $$\max_{\underline{s}_i \in \mathcal{S}_i} \| E_i(\underline{s}_i) \| < 1 \tag{7.3d}$$

In terms of (7.1)-(7.3) the composite system can be described as :

$$\underline{x}(k+1) = [A + \Delta A(\underline{r}) + M]\, \underline{x}(k) + [B + \Delta B(\underline{s})]\, \underline{u}(k) + C\, \underline{v}(k) + $$

$$+ H(k,\underline{x}(k),\underline{r}(k)) \qquad (7.4a)$$

$$\underline{x}(0) = \underline{x}_0 \qquad (7.4b)$$

$$\underline{y}(k) = \underline{x}(k) + \underline{w}(k) \qquad (7.5)$$

where $\underline{x} = [\underline{x}_1^T, \underline{x}_2^T, \ldots\ldots, \underline{x}_N^T]^T \in R^n$ is the composite state ; $n = \sum_{i=1}^{N} n_i$, $\underline{u} = [\underline{u}_1^T, \underline{u}_2^T, \ldots\ldots, \underline{u}_N^T]^T \in R^m$ is the composite control ; $m = \sum_{i=1}^{N} m_i$, $\underline{v} = [\underline{v}_1^T, \underline{v}_2^T, \ldots\ldots, \underline{v}_N^T]^T \in R^l$ is the composite disturbance ; $l = \sum_{i=1}^{N} l_i$, $\underline{y} = [\underline{y}_1^T, \underline{y}_2^T, \ldots\ldots, \underline{y}_N^T]^T \in R^n$ is the composite measured state , $\underline{w} = [\underline{w}_1^T, \underline{w}_2^T, \ldots, \underline{w}_N^T]^T \in R^n$ is the composite measurement error , M and H(.,.,.) represent respectively the interactions and their uncertainties of the overall system. Furthermore , $A = \text{diag}(A_i)$, $B = \text{diag}(B_i)$, $C = \text{diag}(C_i)$, $\Delta A(.) = \text{diag}(\Delta A_i(.))$ and $\Delta B(.) = \text{diag}(\Delta B_i(.))$. It is interesting to note that both M and H are off-diagonal matrices , i.e. $M_{ij} = H_{ij} = 0$ for $i \neq j$.

Based on the scenario above, our goal is to design large scale control system methodologies in order to stabilize the interconnected uncertain dynamic system at hand. Before doing that, we have to recall preliminary results of ultimate boundedness theory [38-45] :

Definition 7.2.1 (Practical Stabilizability)

The uncertain composite dynamical system (7.4) is said to be practically stabilizable if , given any $\underline{d} > 0$, there is a control law $\underline{g}(.,.) : R^n \times \mathcal{Z} \longrightarrow R^m$ for which, given any admissible uncertainties $\underline{r}(.) \in \mathcal{R}$, $\underline{s}(.) \in \mathcal{S}$, $\underline{v}(.) \in \mathcal{V}$ and $\underline{w}(.) \in \mathcal{W}$ (where $\mathcal{R} = \bigcup_{i=1}^{N} \mathcal{R}_i$, $\mathcal{S} = \bigcup_{i=1}^{N} \mathcal{S}_i$ and so on), any initial time $k_0 \in \mathcal{Z}$ and any initial state $\underline{x}_0 \in R^n$, the following conditions hold :

(i) The closed loop system

$$\underline{x}(k+1) = [\tilde{A} + \Delta A(\underline{r}) + M]\, \underline{x}(k) + [B + \Delta B(\underline{s})]\, \underline{g}(\underline{y}(k),k) + C\, \underline{v}(k)$$

$$+ H(k,\underline{r}(k),\underline{x}(k)) \qquad (7.6)$$

possesses a solution $\underline{x}(.) : [k_0,k_1] \longrightarrow R^n$, $\underline{x}(k_0) = \underline{x}_0$

(ii) Given any $\nu > 0$ and any solution $\underline{x}(.) : [k_0,k_1] \longrightarrow R^n$, $\underline{x}(k_0) = \underline{x}_0$ of (7.6) with $\| \underline{x}_0 \| \le \nu$, there is a constant $d(\nu) > 0$ such that $\| \underline{x}(k) \| \le d(\nu) \quad \forall \quad k \in [k_0,k_1]$.

(iii) Every solution $\underline{x}(.) : [k_0,k_1] \longrightarrow R^n$ can be continued over $[k_0,\infty)$.

(iv) Given any $\bar{d} \ge \underline{d}$, any $\nu > 0$ and any solution $\underline{x}(.) : [k_0,\infty) \longrightarrow R^n$, $\underline{x}(k_0) = \underline{x}_0$ of (7.6) with $\| \underline{x}_0 \| \le \nu$, there exists a finite period $\tilde{K}(\bar{d},\nu) < \infty$, possibly dependent on ν but not on k_0, such that $\| \underline{x}(k) \| \le \bar{d} \quad \forall \quad k \ge k_0 + \tilde{K}(\bar{d},\nu)$.

(v) Given any $\bar{d} \ge \underline{d}$ and any solution $\underline{x}(.) : [k_0,k_1] \longrightarrow R^n$, $\underline{x}(k_0) = \underline{x}_0$, of (7.6), there is a constant $\delta(\bar{d}) > 0$ such that $\| \underline{x}(k_0) \| \le \delta(\bar{d})$ implies $\| \underline{x}(k) \| \le \bar{d} \quad \forall \quad k \ge k_0$.

7.3 Stabilization of Decoupled Subsystems

For generality, we assume that the subsystem matrix A_i is unstable. Now, given a matrix G_i satisfying Assumption (A4), consider the class of decentralized feedback controls :

$$\underline{u}_i(k) = G_i \; \underline{y}_i(k) + \underline{g}_i(\underline{y}_i(k),k) \qquad \forall \quad \underline{y}_i \in R^n \tag{7.7}$$

where $\underline{g}_i(.,.) : R^n \longrightarrow R^m$ is given by

$$\underline{g}_i(\underline{y}_i,k) = \begin{cases} \dfrac{-B_i^T P_i \tilde{A}_i \underline{y}_i}{\| B_i^T P_i \tilde{A}_i \underline{y}_i \|} \sigma_i(\underline{y}_i) & \text{for } \|B_i^T P_i \tilde{A}_i \underline{y}_i\| > \epsilon_i \\[2ex] \dfrac{-B_i^T P_i \tilde{A}_i \underline{y}_i}{\epsilon_i} \sigma_i(\underline{y}_i) & \text{for } \|B_i^T P_i \tilde{A}_i \underline{y}_i\| \le \epsilon_i \end{cases} \tag{7.8}$$

where ϵ_i is a pre-specified positive constant , P_i is a solution of Lyapunov equation :

$$\tilde{A}_i^T P_i \tilde{A}_i - P_i = - Q_i \qquad ; \quad Q_i > 0 \tag{7.9}$$

and $\sigma_i(.) : R^{n_i} \longrightarrow R_+$ is non-negative function chosen to satisfy :

$$\begin{aligned} \sigma_i(\underline{y}_i) = [\, 1 - \max_{\underline{s}_i \in S_i} \| E_i(\underline{s}_i) \| \,]^{-1} \{ & \max_{\underline{r}_i \in \mathcal{R}_i} \| D_i(\underline{r}_i)\, \underline{y}_i \| \\ & + \max_{\substack{\underline{r}_i \in \mathcal{R}_i \\ \underline{w}_i \in \mathcal{W}_i}} \| D_i(\underline{r}_i)\, \underline{w}_i \| + \max_{\underline{w}_i \in \mathcal{W}_i} \| G_i \, \underline{w}_i \| \\ & + \max_{\underline{s}_i \in S_i} \| E_i(\underline{s}_i)\, G_i \, \underline{y}_i \| + \max_{\underline{v}_i \in \mathcal{V}_i} \| F_i \, \underline{v}_i \| \} \end{aligned} \tag{7.10}$$

For ease of exposition , we define the following norm quantities :

$$\rho_{ri} = \underset{\underline{r}_i \in \mathcal{R}_i}{\text{Max}} \| D_i(\underline{r}_i) \| \quad ; \quad \rho_{si} = \underset{\underline{s}_i \in S_i}{\text{Max}} \| E_i(\underline{s}_i) \| \tag{7.11a}$$

$$\rho_{vi} = \underset{\underline{v}_i \in \mathcal{V}_i}{\text{Max}} \| F_i \underline{v}_i \| \quad ; \quad \rho_{wi} = \underset{\underline{w}_i \in \mathcal{W}_i}{\text{Max}} \| \underline{w}_i(k) \| \tag{7.11b}$$

$$\rho_{gi} = \| G_i \| \quad ; \quad \rho_{sgi} = \underset{\underline{s}_i \in S_i}{\text{Max}} \| E_i(\underline{s}_i) G_i \| \tag{7.11c}$$

In view of (7.2), (7.10) and (7.11), it is sraightforward to see that

$$\sigma_i(\underline{y}_i) \le \sigma_i(\underline{x}_i)$$

$$\triangleq a_i + b_i \| \underline{x}_i \| \tag{7.12}$$

where

$$a_i = [(2 \rho_{ri} + \rho_{gi} + \rho_{sgi}) \rho_{wi} + \rho_{vi}] / (1 - \rho_{si}) \tag{7.13a}$$

$$b_i = (\rho_{ri} + \rho_{sgi}) / (1 - \rho_{si}) \tag{7.13b}$$

The following additional assumption is concerned with the boundedness of the interactions between the subsystems :
(A6) The interactions $H_{ij}(k, \underline{r}_i, \underline{x}_j)$ are assumed to satisfy

$$\| \sum_{\substack{j=1 \\ \neq i}}^{N} H_{ij}(k, \underline{r}_i, \underline{x}_j) \| \le \sum_{\substack{j=1 \\ \neq i}}^{N} \| \Delta A_{ij}(k, \underline{r}_i) \| . \| \underline{x}_j \|$$

$$\le \sum_{\substack{j=1 \\ \neq i}}^{N} \gamma_{ij} \| \underline{x}_j \| \quad \forall \ (k, \underline{r}_i, \underline{x}_j) \in \mathcal{Z} \times R^{p_i} \times R^{n_i} \ ;$$

$$; \quad i=1,2,\ldots\ldots, N \tag{7.14}$$

where γ_{ij} are N^2 non-negative upper bounds for the uncertainties among the subsystems.

The following theorem sets up the condition for decentralized stabilization with respect to the strength (the magnitude of information flow) of the interactions as well as the interactions' uncertainties among various subsystems.

Theorem 7.4.1

The composite system (7.4) satisfying Assumptions (A1)-(A6) can be practically stabilized in a decentralized fashion by the local controls (7.7) if the test matrix given by $L = [l_{ij}]$;

$$l_{ij} = \begin{cases} \lambda_m (Q_i) & \text{for } i = j \\ -\lambda_M (P_i) \left[\|A_{ij}\|^2 + (\|A_{ij}\| + \|A_{ji}\|)\gamma + \gamma^2 \right) \\ - \|P_i \tilde{A}_i\| \left[\|A_{ij}\| + \|A_{ji}\| + 2\gamma\right] & \text{for } i \neq j \end{cases} \tag{7.15}$$

is positive definite matrix.
where γ , represents the total bound for interactions' uncertainty , is defined as :

$$\gamma = \sum_{i=1}^{N} \sum_{j=1}^{N} \gamma_{ij} . \tag{7.16}$$

Proof

The system (7.1) under the application of the control (7.7) and utilizing the Assumptions (A1)-(A6) can be written as :

$$\underline{x}_i(k+1) = A_i \underline{x}_i(k) + B_i \underline{g}_i(\underline{y}_i(k),k) + B_i \underline{\Phi}_i(\underline{x}_i(k),k) + \sum_{\substack{j=1 \\ j\neq i}}^{N} A_{ij} \underline{x}_j(k) + \sum_{\substack{j=1 \\ j\neq i}}^{N} H_{ij}(k,\underline{r}_i,\underline{x}_j(k)) \tag{7.17}$$

where

$$\underline{\Phi}_i(\underline{x}_i, k) = D_i(\underline{r}_i)\,\underline{x}_i + G_i\,\underline{w}_i + E_i(\underline{s}_i)\,G_i\,(\underline{x}_i + \underline{w}_i) +$$

$$+ E_i(\underline{s}_i)\,\underline{g}_i(\underline{x}_i + \underline{w}_i, k) + F_i\,\underline{v}_i \qquad (7.18)$$

In view of (7.8) and (7.10), one can easily show that :

$$\| \underline{\Phi}_i(\underline{x}_i, k) \| \le \sigma_i(\underline{y}_i) \qquad (7.19)$$

Now , let the Lyapunov function V_i be as an index of " energy " associated with $\underline{x}_i$ and is chosen as

$$V_i(\underline{x}_i(k), k) = \underline{x}_i^T\,P_i\,\underline{x}_i \qquad ; \quad i=1,2,\ldots\ldots,N \qquad (7.20)$$

where P_i is the solution of (7.9). Then it is tempting to define

$$V(k) = \sum_{i=1}^{N} V_i \qquad (7.21)$$

as the energy of the composite system. Taking the forward difference of (7.20), one gets :

$$\Delta V_i(k) = \underline{x}_i^T(k+1)\,P_i\,\underline{x}_i(k+1) - \underline{x}_i^T(k)\,P_i\,\underline{x}_i(k) \qquad (7.22)$$

A little algebra on (7.22) using (7.2),(7.8),(7.9),(7.19) and dropping the suffices for simplicity, yields :

$$\Delta V_i = \underline{x}_i^T\,[\tilde{A}_i^T\,P_i\,\tilde{A}_i - P_i\,]\underline{x}_i + 2\,\underline{x}_i^T\,\tilde{A}_i^T\,P_i\,B_i\,[\underline{g}_i + \underline{\Phi}_i] +$$

$$+ 2\,\underline{x}_i^T\,\tilde{A}_i^T\,P_i \sum_{j=1}^{N} (A_{ij}\,\underline{x}_j + H_{ij}) + [\underline{g}_i + \underline{\Phi}_i]^T\,B_i^T\,P_i\,B_i\,\bullet$$

$$\cdot [\underline{g}_i + \underline{\Phi}_i] + 2(\underline{g}_i + \underline{\Phi}_i)^T B_i^T P_i \sum_{j=1}^{N} (A_{ij}\underline{x}_j + H_{ij})$$

$$+ [\sum_{j=1}^{N} (A_{ij}\underline{x}_j + H_{ij})]^T P_i [\sum_{j=1}^{N} (A_{ij}\underline{x}_j + H_{ij})]$$

$$\therefore \Delta V_i(k) \le -\underline{x}_i^T Q_i \underline{x}_i + 2(B_i^T P_i \tilde{A}_i \underline{y}_i)^T [\underline{g}_i(\underline{y}_i, k) +$$

$$+ \sigma_i B_i^T P_i \tilde{A}_i \underline{y}_i / \|B_i^T P_i \tilde{A}_i \underline{y}_i\|] - 2(B_i^T P_i \tilde{A}_i \underline{w}_i)\cdot$$

$$\cdot [\underline{g}_i(\underline{y}_i, k) - \sigma_i(\underline{y}_i) B_i^T P_i \tilde{A}_i \underline{w}_i / \|B_i^T P_i \tilde{A}_i \underline{w}_i\|]$$

$$+ 2\underline{x}_i^T \tilde{A}_i^T P_i \sum_{j=1}^{N} (A_{ij} + \gamma_{ij})\underline{x}_j +$$

$$+ [\underline{g}_i(\underline{y}_i, k) + \sigma_i B_i^T P_i \tilde{A}_i \underline{y}_i / \|B_i^T P_i \tilde{A}_i \underline{y}_i\|]^T B_i^T P_i B_i \cdot$$

$$\cdot [\underline{g}_i(\underline{y}_i, k) + \sigma_i B_i^T P_i \tilde{A}_i \underline{y}_i / \|B_i^T P_i \tilde{A}_i \underline{y}_i\|] + 2[\underline{g}_i(\underline{y}_i, k)$$

$$+ \sigma_i B_i^T P_i \tilde{A}_i \underline{y}_i / \|B_i^T P_i \tilde{A}_i \underline{y}_i\|]^T B_i^T P_i \sum_{j=1}^{N} (A_{ij} + \gamma_{ij})\underline{x}_j +$$

$$+ [\sum_{j=1}^{N} (A_{ij} + \gamma_{ij})\underline{x}_j]^T P_i [\sum_{j=1}^{N} (A_{ij} + \gamma_{ij})\underline{x}_j] \quad (7.23)$$

As a consequence of (7.8) and (7.16), the second term on the r.h.s. of (7.23) vanishes for $\|B_i^T P_i \tilde{A}_i \underline{y}_i\| > \epsilon_i$, but if $\|B_i^T P_i \tilde{A}_i \underline{y}_i\| \le \epsilon_i$, its maximum value (at $\|B_i^T P_i \tilde{A}_i \underline{y}_i\| = \epsilon_i/2$) is equal to $\epsilon_i \sigma_i(\underline{y}_i)/2$. The maximum value of the third term occurring for $\|B_i^T P_i \tilde{A}_i \underline{y}_i\| > \epsilon_i$ and $B_i^T P_i \tilde{A}_i \underline{y}_i / \|B_i^T P_i \tilde{A}_i \underline{y}_i\| = B_i^T P_i \tilde{A}_i \underline{w}_i / \|B_i^T P_i \tilde{A}_i \underline{w}_i\|$ is simply $4\|B_i^T P_i \tilde{A}_i \underline{w}_i\| \sigma_i(\underline{y}_i)$. The maximum value of the fourth term is equal to

$2\,\|P_i\,\tilde{A}_i\|\,\sum_{j=1}^{N}(\|A_{ij}\|+\gamma)\|\underline{x}_j\|^2$. With respect to the fifth and sixth terms, they vanish for $\|B_i^T\,P_i\,\tilde{A}_i\,\underline{y}_i\| > \epsilon_i$ but if $\|B_i^T\,P_i\,\tilde{A}_i\,\underline{y}_i\| \le \epsilon_i$, their maximum values occur at $\|B_i^T\,P_i\,\tilde{A}_i\,\underline{y}_i\| = \epsilon_i$ lest the terms vanish again. The maximum value of the last term is equal to $2\,\lambda_M(P_i)\,\sum_{j=1}^{N}(\|A_{ij}\|+\gamma)^2\,\|\underline{x}_j\|^2$. Regrouping the terms and summing up the maxima , we have :

$$\Delta V_i \le -\,\underline{x}_i^T\,Q_i\,\underline{x}_i + [\,\epsilon_i/2 + 4\,\|B_i^T\,P_i\,\tilde{A}_i\,\underline{w}_i\|\,]\,\sigma_i(\underline{y}_i) + $$

$$+\,2\,\|P_i\,\tilde{A}_i\|\,\sum_{j=1}^{N}(\|A_{ij}\|+\gamma)\|\underline{x}_j\|^2 +$$

$$+\,2\,\lambda_M(P_i)\,\sum_{j=1}^{N}(\|A_{ij}\|+\gamma)^2\,\|\underline{x}_j\|^2 \tag{7.24}$$

$$\therefore\ \Delta V(k) \le -\,\underline{x}^T\,L\,\underline{x} + \theta_1\,\|\underline{x}\| + \theta_0 \tag{7.25}$$

where the NxN symmetric matrix L is given by (7.15) and

$$\theta_0 = \sum_{i=1}^{N} a_i\,(\,\epsilon_i/2 + 4\,\|B_i^T\,P_i\,\tilde{A}_i\|\,\rho_{wi}\,) \tag{7.26a}$$

$$\theta_1 = \sum_{i=1}^{N} b_i\,(\,\epsilon_i/2 + 4\,\|B_i^T\,P_i\,\tilde{A}_i\|\,\rho_{wi}\,) \tag{7.26b}$$

Now , ΔV is negative definite if the test matrix L is positive definite for all $(\underline{x},k) \in \Omega^c(\eta) \times \mathcal{Z}$ where $\Omega^c(\eta)$ is the complement of the set $\Omega(\eta)$, whereas $\Omega(\eta)$ is a closed sphere with radius η given by

$$\eta = \{ \theta_1 + \sqrt{\theta_1^2 + 4\,\theta_0\,\lambda_m(L)}\ \} / (2\,\lambda_m(L)) \tag{7.27}$$

Therefore, in view of the theory of ultimate boundedness [38-45], it suffices to get $\underline{d} > 0$ in Definition 7.2.1. We define it as the major axis of the smallest ellipsoid containing $\Omega(\eta)$, using the standard arguments in [38-45], $\underline{d}$ will be given by :

$$\underline{d} = \eta \sqrt{\lambda_M(P) / \lambda_m(P)} \tag{7.28}$$

where $P = \text{diag}(P_i)$ (7.29)

This concludes the proof of Theorem 7.4.1. ■■■

Corollary 7.4.1

In the absence of state uncertainty ; that is , $\underline{w}_i = \underline{0}$, and if there are infinite feedback gains ; that is , $\epsilon_i = 0$,then the nonlinear (switching) term of the controller (7.7) will be given by :

$$\underline{g}_i(\underline{y}_i,k) = \underline{g}_i(\underline{x}_i,k) =$$

$$= \begin{cases} -\sigma_i(\underline{x}_i)(B_i^T P_i \tilde{A}_i \underline{x}_i)/\|B_i^T P_i \tilde{A}_i \underline{x}_i\| & \text{for } \|B_i^T P_i \tilde{A}_i \underline{x}_i\| \neq 0 \\ \epsilon_i \{ \underline{\xi}_i \in R^n : \|\underline{\xi}_i\| \le \sigma_i(\underline{x}_i) & \text{for } \|B_i^T P_i \tilde{A}_i \underline{x}_i\| = 0 \end{cases} \tag{7.30}$$

It follows from (7.27) in this case that $\eta = 0$ and Ω = empty set ; that is , Ω^c will represent the whole space. On the other hand from (7.25) , we find that

$$\Delta V \le -\underline{x}^T L\, \underline{x} \qquad \forall\ (\underline{x},k) \in R^n \times \mathcal{Z} \tag{7.31}$$

7.5 Hierarchical Control Structure

By adopting the concept of hierarchical control, the control signal is decomposed into two-components. The first one, generated by a higher level controller (coordinator), balances completely the effect of interactions between the subsystems. The second one, generated by local controllers, gives the desired boundedness behaviour for the decoupled uncertain subsystems.

Now, given a matrix G_i satisfying (A4), consider the class of hierarchical feedback controls :

$$\underline{u}(k) = G_b\ \underline{y}(k) + \underline{g}(\underline{y}(k),k) \tag{7.32}$$

where G_b = diag (G_i) and $\underline{g}(.,.) : R^n \longrightarrow R^m$ is given by

$$\underline{g}(\underline{y}(k),k) = \begin{cases} -\sigma(\underline{y})\ B^T\ P\ \tilde{A}\ \underline{y}/|B^T\ P\ \tilde{A}\ \underline{y}| & \text{for } \|B^T\ P\ \tilde{A}\ \underline{y}\| > \epsilon \\ -\sigma(\underline{y})\ B^T\ P\ \tilde{A}\ \underline{y}/\epsilon & \text{for } \|B^T\ P\ \tilde{A}\ \underline{y}\| \le \epsilon \end{cases} \tag{7.33}$$

where ϵ is a pre-specified positive constant , $\tilde{A}$ = diag ($\tilde{A}_i$) ; P = diag (P_i); P_i is the solution of (7.9).

and $\sigma(.) : R^n \longrightarrow R_+$ is non-negative function chosen to satisfy :

$$\begin{aligned} \sigma(\underline{y}) = \underset{i}{\mathrm{Max}} \{ [1 - \underset{\underline{s}_i \in S_i}{\mathrm{Max}} \|E_i(\underline{s}_i)\|]^{-1} [\underset{\underline{r}_i \in \mathcal{R}_i}{\mathrm{Max}} \|D_i(\underline{r}_i)\ \underline{y}_i\| + \\ + \underset{\substack{\underline{r}_i \in \mathcal{R}_i \\ \underline{w}_i \in \mathcal{W}_i}}{\mathrm{Max}} \|D_i(\underline{r}_i)\ \underline{w}_i\| + \underset{\underline{w}_i \in \mathcal{W}_i}{\mathrm{Max}} \|G_i\ \underline{w}_i\| + \\ + \underset{\underline{s}_i \in S_i}{\mathrm{Max}} \|E_i(\underline{s}_i)\ G_i\ \underline{y}_i\| + \underset{\underline{v}_i \in \mathcal{V}_i}{\mathrm{Max}} \|F_i\ \underline{v}_i\| \} \\ ; \quad i=1,2,......,N \end{aligned} \tag{7.34}$$

In view of (7.5), (7.11) and (7.34), we can find that :

$$\sigma(\underline{y}) \le \sigma(\underline{x})$$

$$= \tilde{a} + \tilde{b}\,\|\underline{x}\| \qquad (7.35)$$

where

$$\tilde{a} = \max_i \{[(2\rho_{ri} + \rho_{gi} + \rho_{sgi})\rho_{wi} + \rho_{vi}]/(1 - \rho_{si})\} \qquad (7.36a)$$

$$\tilde{b} = \max_i \{[\rho_{ri} + \rho_{sgi}]/(1 - \rho_{si})\} \qquad (7.36b)$$

where i=1,2,......,N

Theorem 7.5.1

The composite system (7.4), (7.5) satisfying (A1)-(A7) can be practically stabilized via the hierarchical control (7.32) if the test matrix L given by (7.15) is positive definite. Furthermore, the resulting closed - loop state trajectories are bounded in a set $\Omega^c(\bar{\eta})$ where $\Omega^c(\bar{\eta})$ is the complement of a set $\Omega(\bar{\eta})$ with radius $\bar{\eta}$ given by

$$\bar{\eta} = [\mu_1 + \sqrt{\mu_1^2 + 4\mu_0\lambda_m(L)}]\,/\,[2\lambda_m(L)] \qquad (7.37)$$

where

$$\mu_0 = \tilde{a}\,(\epsilon/2 + 4\,\|B^T P\tilde{A}\|\,\rho_w) \qquad (7.38a)$$

$$\mu_1 = \tilde{b}\,(\epsilon/2 + 4\,\|B^T P\tilde{A}\|\,\rho_w) \qquad (7.38b)$$

$$\rho_w = \max_i \{\rho_{wi}\} \quad ; \quad i=1,2,\ldots,N \qquad (7.38c)$$

and $\tilde{a}$, $\tilde{b}$ are given by (7.36).

Proof

The closed loop system (7.4) with the control law (7.32) becomes :

$$\underline{x}(k+1) = (\tilde{A} + M)\underline{x}(k) + B[\underline{g}(\underline{y},k) + \underline{\Phi}(\underline{x},k)] + H(k,\underline{r},\underline{x}) \qquad (7.39)$$

where

$$\underline{\Phi}(x,k) = D(\underline{r})\ \underline{x} + E(\underline{s})\ G\ (\underline{x} + \underline{w}) + E(\underline{s})\ \underline{g}(\underline{y},k) + F\ \underline{v} \qquad (7.40)$$

In view of (7.11),(7.34) and (7.35), one can obtain :

$$\|\ \underline{\Phi}(\underline{x},k)\ \| \le \sigma(\underline{y}) \le \sigma(\underline{x})$$

$$\triangleq \tilde{a} + \tilde{b}\ \|\ \underline{x}\ \| \qquad (7.41)$$

where $\tilde{a}$ and $\tilde{b}$ are given by (7.36). Now , defining a candidate Lyapunov function, V(k), as

$$V(k) = \sum_{i=1}^{N} \underline{x}_i^T(k)\ P_i\ \underline{x}_i(k) \qquad (7.42)$$

where P_i is the solution of (7.9). Taking the forward difference of V(k), we have :

$$\Delta V(k) = \sum_{i=1}^{N} \{\ \underline{x}_i^T(k+1)\ P_i\ \underline{x}_i(k+1) - \underline{x}_i^T(k)\ P_i\ \underline{x}_i(k)\}$$

$$= \sum_{i=1}^{N} (\{\tilde{A}_i\underline{x}_i(k) + B_i[\underline{g}_i(\underline{y}_i,k) + \underline{\Phi}_i(\underline{x}_i,k)] + \underline{\mathfrak{z}}_i + \underline{h}_i\}^T\ P_i\ \cdot$$

$$\cdot\{\tilde{A}_i\underline{x}_i(k) + B_i[\underline{g}_i(\underline{y}_i,k) + \underline{\Phi}_i(\underline{x}_i,k)] + \underline{\mathfrak{z}}_i + \underline{h}_i\}$$

$$- \underline{x}_i^T(k)\ P_i\ \underline{x}_i(k)\ ) \qquad (7.43)$$

A little algebra on (7.43) using (7.2), (7.9), (7.32), (7.33), (7.41) and dropping the suffices for simplicity, yields :

$$\Delta V(k) \leq - \underline{x}^T \, Q \, \underline{x} + [\epsilon/2 + 4 \, \| B^T \, P \, \tilde{A} \, \underline{w} \| \,] \, \sigma(\underline{y}) +$$

$$+ \, 2 \sum_{i=1}^{N} \| P_i \, \tilde{A}_i \| \sum_{j=1}^{N} (\| A_{ij} \| + \gamma) \, \| \underline{x}_j \|^2 +$$

$$+ \, 2 \sum_{i=1}^{N} \lambda_M(P_i) \sum_{j=1}^{N} (\| A_{ij} \| + \gamma)^2 \, \| \underline{x}_j \|^2 \qquad (7.44)$$

where $Q = \mathrm{diag}(Q_i)$ and $\tilde{A} = \mathrm{diag}(\tilde{A}_i)$. Using (7.11), (7.34)-(7.36), (7.44) can be simplified to :

$$\Delta V(k) \leq - \underline{x}^T(k) \, L \, \underline{x}(k) + \mu_1 \, \| \underline{x}(k) \| + \mu_0 \qquad (7.45)$$

where the NxN symmetric matrix L is given by (7.15) and μ_0, μ_1 are given by (7.38).

Now, $\Delta V(k)$ is negative definite if the test matrix L is positive definite for all $(\underline{x},k) \in \Omega^c(\bar{\eta}) \times \mathcal{Z}$,where $\bar{\eta}$ is given by (7.37). In view of the theory of ultimate boundedness [38-45],it suffices to get $\underline{d} > 0$ in Definition 7.2.1. We define it as the major axis of the smallest ellipsoid containing $\Omega(\bar{\eta})$. Using the standard arguments in [38-45], d will be given by

$$\underline{d} = \bar{\eta} \sqrt{\lambda_M(P) / \lambda_m(P)} \qquad (7.46)$$

which completes the proof of Theorem 7.3.1. ▪▪▪

It is worth mentioning that When $\underline{w}_i = \underline{0}$ and $\epsilon = 0$, it follows from (7.33), with the aid of (7.38) that $\bar{\eta} = 0$ and hence $\Omega^c(\bar{\eta})$ will represent the whole space.

7.6 Stability of The System Under Structural Perturbations

The following results investigate the stability of the overall uncertain system when subjected to worst case structural perturbations [95-97].

Theorem 7.6.1

In the absence of the interconnection links between the two levels, the system will be practically stable if the test matrix Γ given by $\Gamma = [\Gamma_{ij}]$

$$\Gamma_{ij} = \begin{cases} \lambda_m(Q_i) - 2\tilde{b}\,\|B_i^T P_i \tilde{A}_i\| - \tilde{b}^2 \lambda_M(B_i^T P_i B_i) & \text{for } i=j \\ -\lambda_M(P_i)\,[\|A_{ij}\|^2 + (\|A_{ij}\| + \|A_{ji}\|)\gamma + \gamma^2] \\ -\|P_i \tilde{A}_i\|\,[\|A_{ij}\| + \|A_{ji}\| + 2\gamma] - \\ -\tilde{b}\|P_i B_i\|\,(\|A_{ij}\| + \gamma) & \text{for } i \neq j \end{cases} \tag{7.47}$$

is positive definite matrix.

Moreover , the resulting closed-loop state trajectories are bounded in a domain $\Omega^c(\eta_1)$ where

$$\eta_1 = \{\bar{\mu}_1 + \sqrt{\bar{\mu}_1^2 + 4\,\bar{\mu}_0\,\lambda_m(\Gamma)}\}\,/\,[2\,\lambda_m(\Gamma)] \tag{7.48}$$

where

$$\bar{\mu}_0 = \tilde{a}^2 \sum_{i=1}^{N} \lambda_M(B_i^T P_i B_i) \tag{7.49a}$$

$$\bar{\mu}_1 = 2\,\tilde{a} \sum_{i=1}^{N} \|B_i^T P_i \tilde{A}_i\| + 2\,\tilde{a}\,\tilde{b} \sum_{i=1}^{N} \lambda_M(B_i^T P_i B_i) + \tilde{a} \sum_{i=1}^{N} \|P_i B_i\| \sum_{j=1}^{N} (\|A_{ij}\| + \gamma) \tag{7.49b}$$

Proof

By putting $\underline{g}(\underline{y}(k),k) = \underline{0}$ in (7.32) and following the analysis pursued from (7.42) to (7.45). ▪▪▪

Theorem 7.6.2

If the decomposed subsystems are completely decoupled , then the global system is practically stable in a domain of attraction $\Omega^c(\eta_2)$ with radius η_2 given by

$$\eta_2 = [\ \mu_1 + \sqrt{\mu_1^2 + 4\ \mu_0\ \lambda_m(Q)}\ ] \ /\ [2\ \lambda_m(Q)] \tag{7.50}$$

where μ_0 and μ_1 are given by (7.38).

Proof

It can be easily obtained by putting $M = H(k,\underline{r},\underline{x}) = 0$ in (7.39) and following analysis similar to (7.42)-(7.45). ▪▪▪

Theorem 7.6.3

When all the links between the coordinator and subsystems and all interactions among the subsystems are cut , then the overall system is practically stable in a domain of attraction $\Omega^c(\eta_3)$ if the following condition holds :

$$\phi = \lambda_m(Q_i) - 2\ \tilde{b}\ \| B_i^T\ P_i\ \tilde{A}_i \| - \tilde{b}^2\ \lambda_M(B_i^T\ P_i\ B_i)$$

$$> 0 \tag{7.51}$$

Moreover, η_3 is given by

$$\eta_3 = [\ \bar{\bar{\mu}}_1 + \sqrt{\bar{\bar{\mu}}_1^2 + 4\ \bar{\mu}_0\ \phi}\ ] \ /\ (2\ \phi) \tag{7.52}$$

where $\bar{\mu}_0$ is given by (7.40a) and

$$\bar{\bar{\mu}}_1 = 2\,\tilde{a} \sum_{i=1}^{N} \| B_i^T P_i \tilde{A}_i \| + 2\,\tilde{a}\,\tilde{b} \sum_{i=1}^{N} \lambda_M(B_i^T P_i B_i) \qquad (7.53)$$

Proof

The Theorem can be proved by setting $\underline{g}(\underline{y}(k),k) = \underline{0}$ in (7.32) and $M = H(k,\underline{r},\underline{x}) = 0$ in (19) and applying the Lyapunov analysis of (7.42) through (7.45).

Remarks

(1) Theorems 7.6.1 and 7.6.2 establish the efficacy of linear control in stabilizing interconnected subsystems and decomposed subsystems respectively.

(2) Again, if $\underline{w}_i = \underline{0}$ and $\epsilon = 0$, it follows from (7.48), (7.50) and (7.52) that $\eta_1 = \eta_2 = \eta_3 = 0$ and hence the domain of stability in the three cases will represent the whole space.

7.7 Mechanical Manipulator Control Example

Consider a two link manipulator described by the block diagram in Figure 7.2. The following equations hold for the open-loop system [98]:

$$\dot{\underline{X}} = \left[\begin{array}{ccc|ccc} -1/\tau_1 & 0.0 & 0.0 & 0.0 & 0.0 & 0.0 \\ 0.0 & 0.0 & 1.0 & 0.0 & 0.0 & 0.0 \\ k_{11} & -k_{11} & (c_1 - k_{12}) & 0.0 & 0.0 & c_2 \\ \hline 0.0 & 0.0 & 0.0 & -1/\tau_2 & 0.0 & 0.0 \\ 0.0 & 0.0 & 0.0 & 0.0 & 0.0 & 1.0 \\ 0.0 & 0.0 & c_3 & k_{21} & -k_{21} & -k_{22} \end{array}\right] \underline{X} + \left[\begin{array}{c|c} 1/\tau_1 & 0 \\ 0 & 0 \\ 0 & 0 \\ \hline 0 & 1/\tau_2 \\ 0 & 0 \\ 0 & 0 \end{array}\right] \underline{U}$$

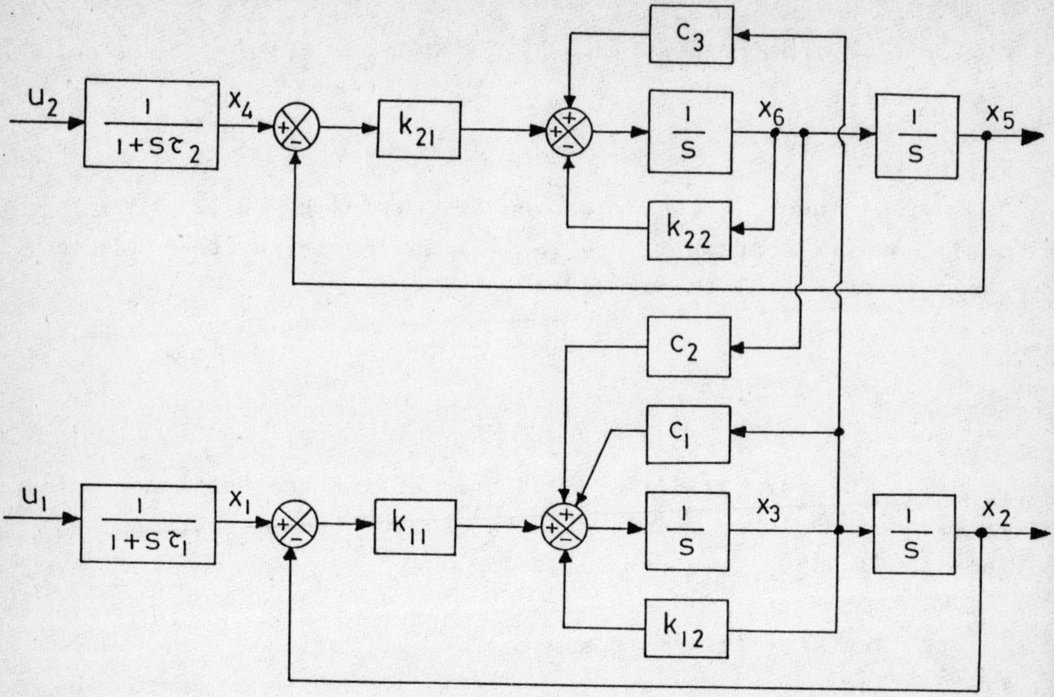

Figure 7.2: A Block Diagram of A Two-link Manipulator Control.

The choice of parameter nominal values with $k_{11}=k_{21}=10$, $k_{12}=k_{22}=2$, $\tau_1=\tau_2=0.1$, $c_1=0.2$, $c_2=c_3=0.1$, together with a discretization scheme with time increment of $\Delta t=0.05$, yields the matrices of system (7.1) of the form :

$$A = \left[\begin{array}{c|c} A_1 & A_{12} \\ \hline A_{21} & A_2 \end{array}\right] = \left[\begin{array}{ccc|ccc} .6065 & 0 & 0 & 0 & 0 & 0 \\ .0103 & .9879 & .0476 & 0 & 0 & .0001 \\ .3731 & -.4762 & .9022 & .001 & -.0012 & .0045 \\ \hline 0 & 0 & 0 & .6065 & 0 & 0 \\ 0 & 0 & .0001 & .0103 & .9879 & .0474 \\ .001 & -.0012 & .0045 & .3711 & -.4738 & .8932 \end{array}\right]$$

$$B^T = \left[\begin{array}{c|c} B_1^T & 0 \\ \hline 0 & B_2^T \end{array}\right] = \left[\begin{array}{ccc|ccc} .3935 & .0018 & .1031 & 0 & 0 & 0 \\ \hline 0 & 0 & 0 & .3935 & .0018 & .1027 \end{array}\right]$$

Suppose now that the coupling coefficients c_1, c_2 and c_3 vary by up to $\pm$ 100% of their nominal values. We thus have :

$$\Delta A(\underline{r}) = \left[\begin{array}{c|c} \Delta A_1(\underline{r}) & 0 \\ \hline 0 & 0 \end{array}\right], \quad \Delta B(s)=0 \text{ and } H(k,\underline{r},\underline{x}) = \left[\begin{array}{c|c} 0 & \Delta A_{12}(\underline{\tilde{r}}) \\ \hline \Delta A_{21}(\underline{\tilde{r}}) & 0 \end{array}\right]$$

where

$$\Delta A_1(\underline{r}) = \begin{bmatrix} 0.0 & 0.0 & 0.0 \\ 0.0 & 0.0 & 0.0 \\ r_1 & r_2 & r_3 \end{bmatrix}$$

$$\Delta A_{12}(\underline{\tilde{r}}) = \Delta A_{21}(\underline{\tilde{r}}) = \begin{bmatrix} 0.0 & 0.0 & 0.0 \\ 0.0 & 0.0 & 0.0 \\ r_4 & r_5 & r_6 \end{bmatrix}$$

and the compact bounding set $\mathcal{R}_1$ is given by

$$\mathcal{R}_1 = \{ \underline{r} \in R^3 : |r_1| \le 0.002, \; |r_2| \le 0.0024, \; |r_3| \le 0.009 \}$$

Moreover, the uncertainties of the interactions are bounded; that is, $|r_4| \le 0.001$, $|r_5| \le 0.0012$ and $|r_6| \le 0.0046$.

It is worth mentioning that the matching conditions are met for the two subsystems with :

$D_1(\underline{r})=0.623[r_1 \quad r_2 \quad r_3]$, $D_2(r)=0$, $E_1(s)=E_2(s)=0$, $F_1=F_2=0$ and hence $\rho_{r1}=0.006$, $\rho_{r2}=\rho_{s1}=\rho_{s2}=\rho_{v1}=\rho_{v2}=\rho_{w1}=\rho_{w2}=0$.

The decentralized control theory [88-92] is applied to compute the gains G_1 and G_2 (with unity weighting matrices) and hence the results are :

$$G_1 = [\ -0.6657 \quad 0.2637 \quad -0.5662\ ] \tag{7.54a}$$

$$G_2 = [\ -0.6536 \quad 0.2465 \quad -0.5447\ ] \tag{7.54b}$$

Thus

$$\tilde{A}_1 \triangleq A_1 + B_1 G_1 = \begin{bmatrix} 0.3445 & 0.1038 & -0.2228 \\ 0.0091 & 0.9884 & 0.0466 \\ 0.3045 & -0.449 & 0.8438 \end{bmatrix}$$

$$\tilde{A}_2 \triangleq A_2 + B_2 G_2 = \begin{bmatrix} 0.3493 & 0.0970 & -0.2143 \\ 0.0091 & 0.9883 & 0.0464 \\ 0.3040 & -0.4485 & 0.8373 \end{bmatrix}$$

With $Q_1 = Q_2 = I_3$, the solution of (7.9), i=1,2 are :

$$P_1 = \begin{bmatrix} 1.6827 & 0.1416 & 0.8989 \\ 0.1416 & 27.5091 & 1.0681 \\ 0.8989 & 1.0681 & 3.0777 \end{bmatrix}$$

$$P_2 = \begin{bmatrix} 1.6785 & 0.1454 & 0.8771 \\ 0.1454 & 27.1675 & 1.0644 \\ 0.8771 & 1.0644 & 3.0124 \end{bmatrix}$$

Proceeding further, we compute

$$\sigma_1(\underline{x}_1) = 0.006\,\|\underline{x}_1\| \quad , \quad \sigma_2(\underline{x}_2) = 0.0$$

With $\epsilon_1 = \epsilon_2 = 0.5$, the nonlinear term is given by :

$$\underline{g}(\underline{x}) = [\; g_1(\underline{x}_1) \;,\; g_2(\underline{x}_2) \;]^T \;, \text{ where for } i = 1,2$$

$$g_i(\underline{x}_i) =$$

$$\begin{cases} -\,\sigma_i(\underline{x}_i) B_i^T P_i \tilde{A}_i \underline{x}_i / \|B_i^T P_i \tilde{A}_i \underline{x}_i\| & \text{for } \|B_i^T P_i \tilde{A}_i \underline{x}_i\| > .5 \\ -\,\sigma_i(\underline{x}_i) B_i^T P_i \tilde{A}_i \underline{x}_i / 0.5 & \text{for } \|B_i^T P_i \tilde{A}_i \underline{x}_i\| \le 0.5 \end{cases} \tag{7.55}$$

(i) The Decentralized Solution

This system was simulated in order to generate the decentralized control structure shown in Figure 7.3.

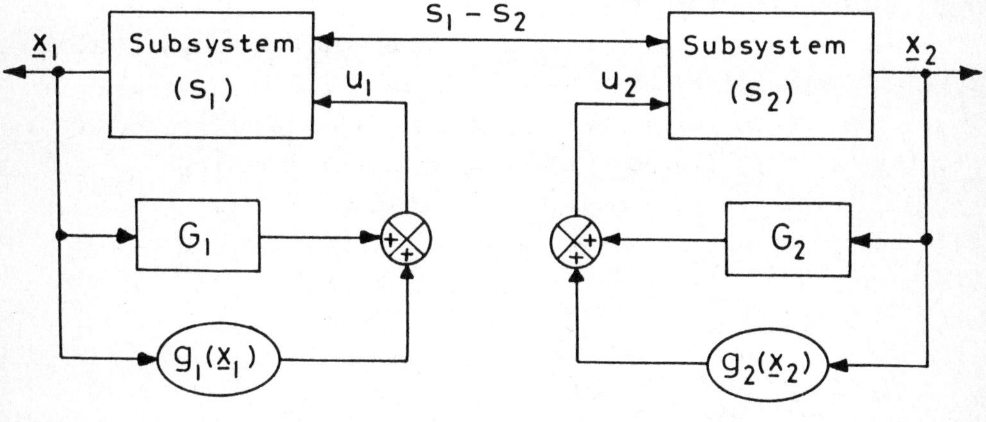

Figure 7.3: The Decentralized Control Structure.

For simulation purposes, the uncertain parameters were taken to be sinusoidal functions of time as follows :

$r_1(k) = 0.0015 \sin(0.25k)$, $r_2(k) = 0.002 \sin(0.01k)$,

$r_3(k) = 0.008 \sin(0.05k)$, $r_4(k) = 0.001 \sin(0.2k)$,

$r_5(k) = 0.001 \sin(0.15k)$ and $r_6(k) = 0.004 \sin(0.1k)$.

Moreover, our system was run for the following two-cases : (a) S_1-S_2 is not cut , (b) S_1-S_2 is cut. and the two cases are shown in Figures (7.4-7.9).

It is worth mentioning that the above problem is solved again by using the centralized technique previously reported in [38-41] and the centralized closed-loop state trajectories are plotted on the same graphs shown in Figures (7.4-7.9). From these simulations, one can conclude that the resulted closed-loop uncertain system in both centralized and decentralized schemes is stable despite the complete absence of liaison between the local control stations in the latter. Moreover, we can see that the trajectories in the case (a) are very close to the centralized ones. This means that the developed decentralized controller provides better performance for the case (a) than for the case (b). However, the centralized control depends in general on the entire state $\underline{X}$ of the global system. Thus, the major advantage of the proposed decentralized design is to reduce the states required for each local control station.

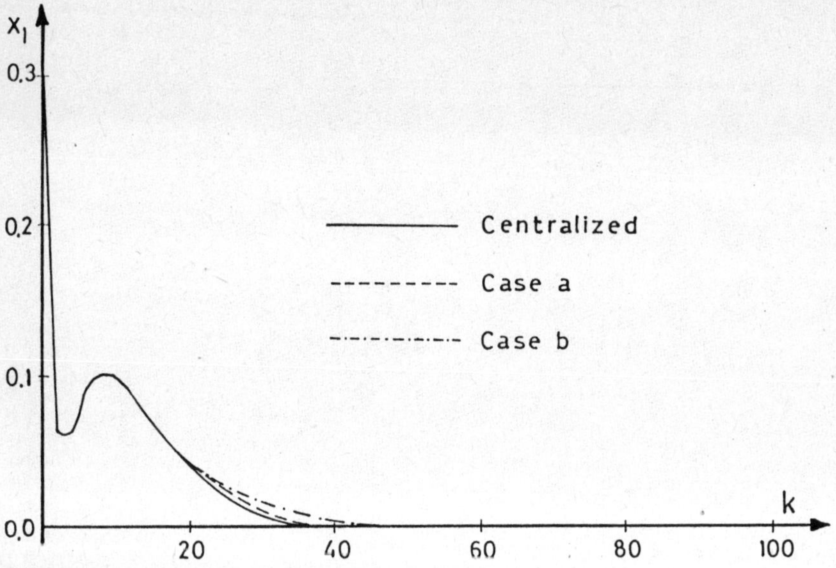

Figure 7.4

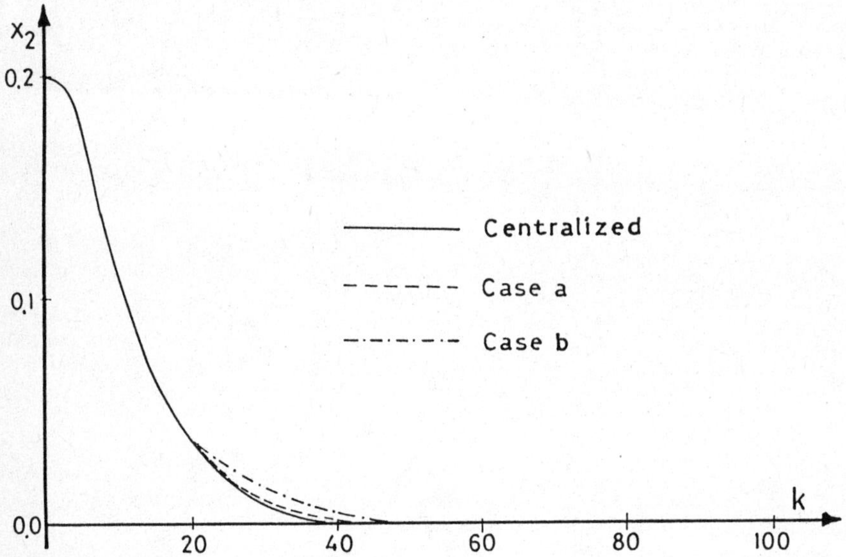

Figure 7.5

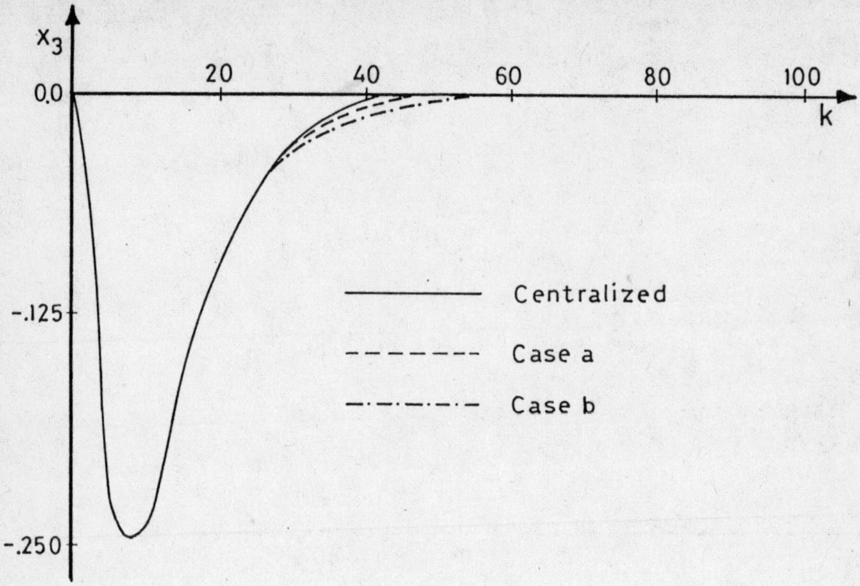

Figure 7.6

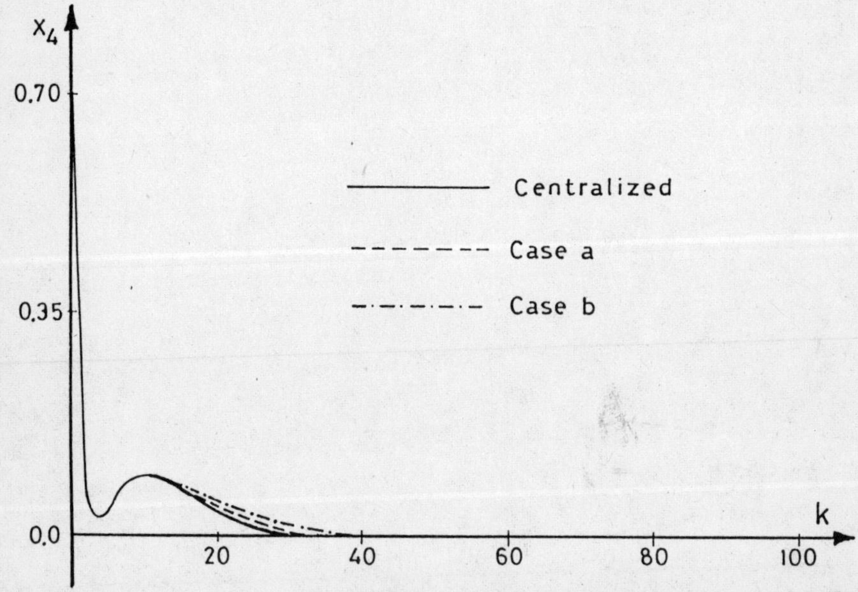

Figure 7.7

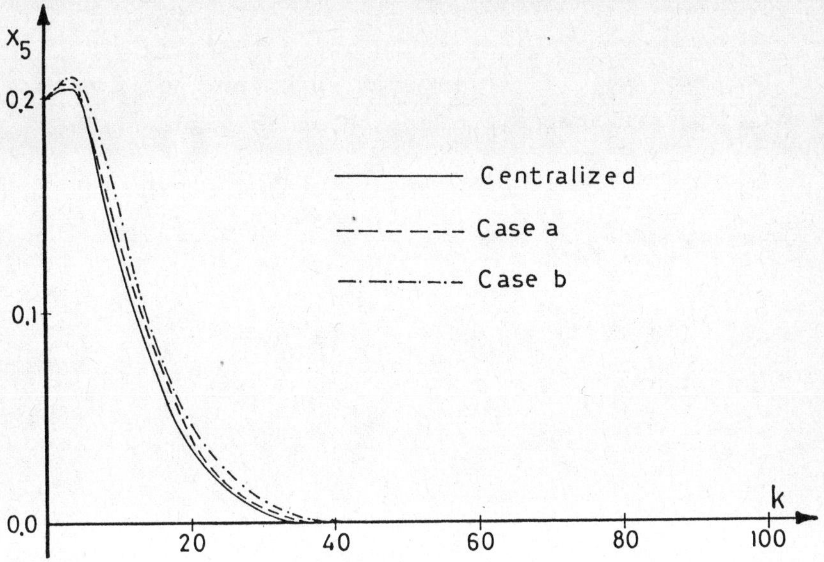

Figure 7.8.

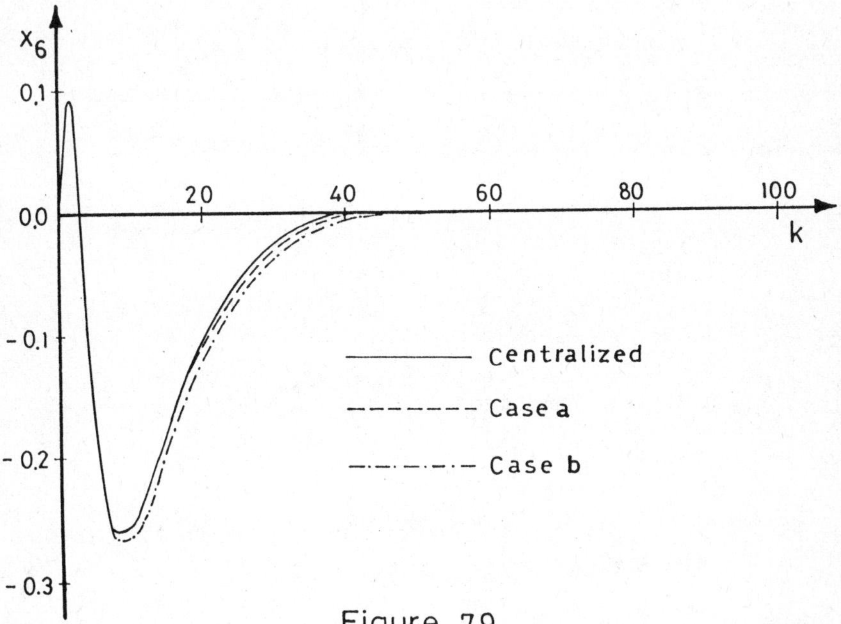

Figure 7.9

(ii) <u>The Hierarchical solution</u>

The problem is solved now in order to generate the following hierarchical structure shown in Figure 7.10.

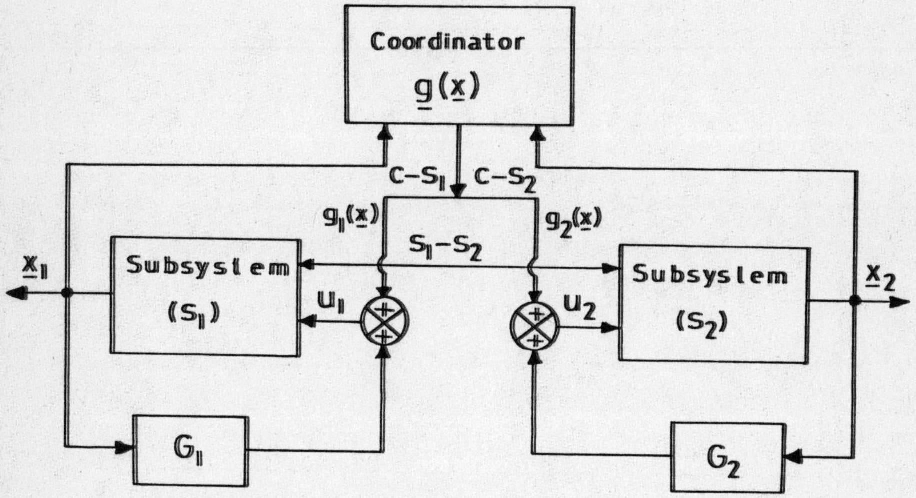

Figure 7.10: The Hierarchical Control Structure.

where G_1 , G_2 are given by (7.54) and $g_1(\underline{x})$, $g_2(\underline{x})$ are given by (7.55) with

$$\sigma_1(\underline{x}) = \sigma_2(\underline{x}) = \sigma(\underline{x}) = 0.006 \,\|\underline{x}\| \quad ; \quad \underline{x} = (\underline{x}_1^T \;,\; \underline{x}_2^T)^T .$$

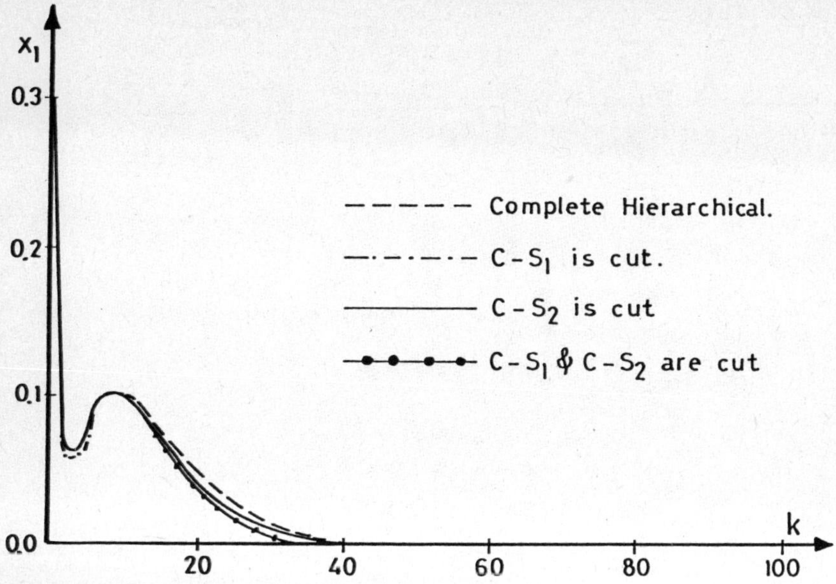

Figure 7.11

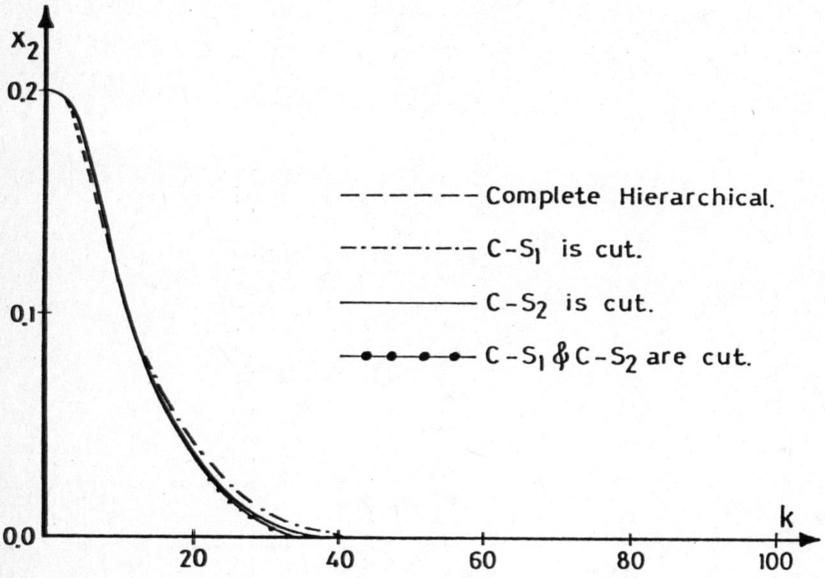

Figure 7.12

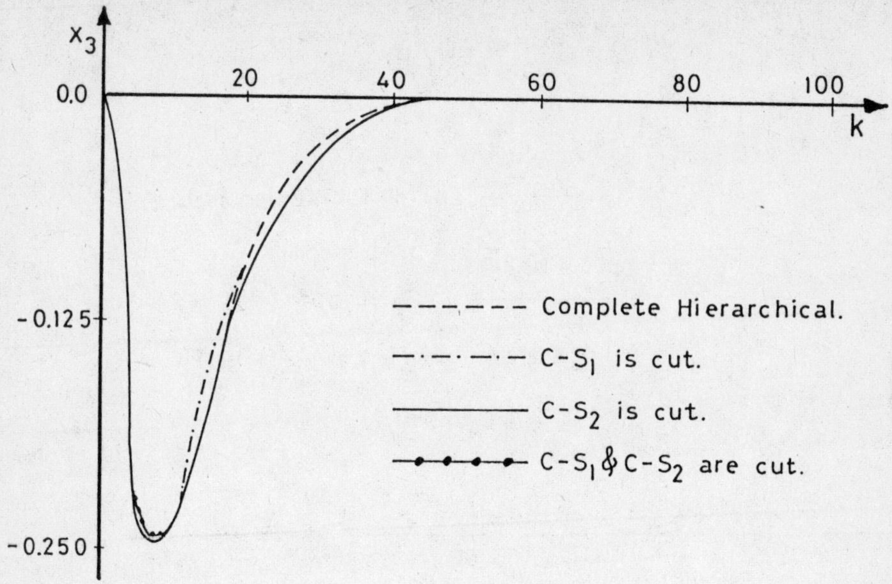

Figure 7.13

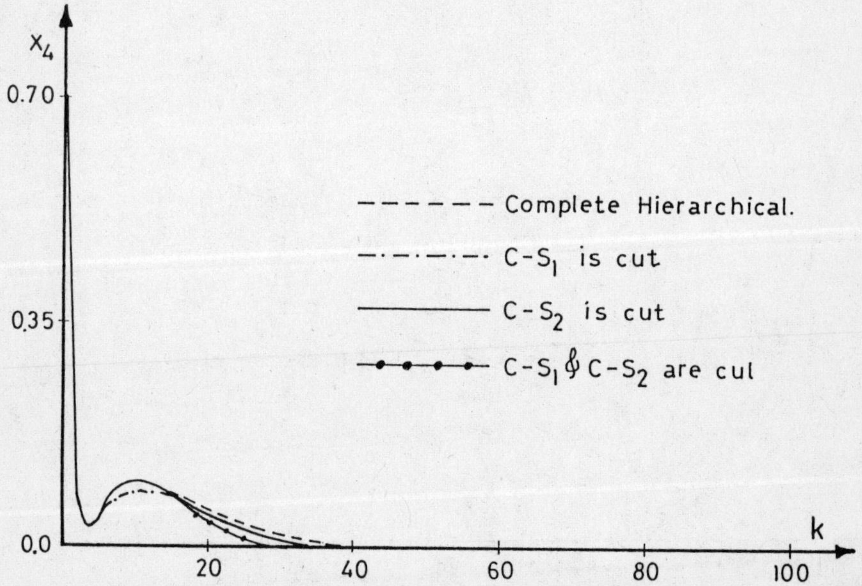

Figure 7.14

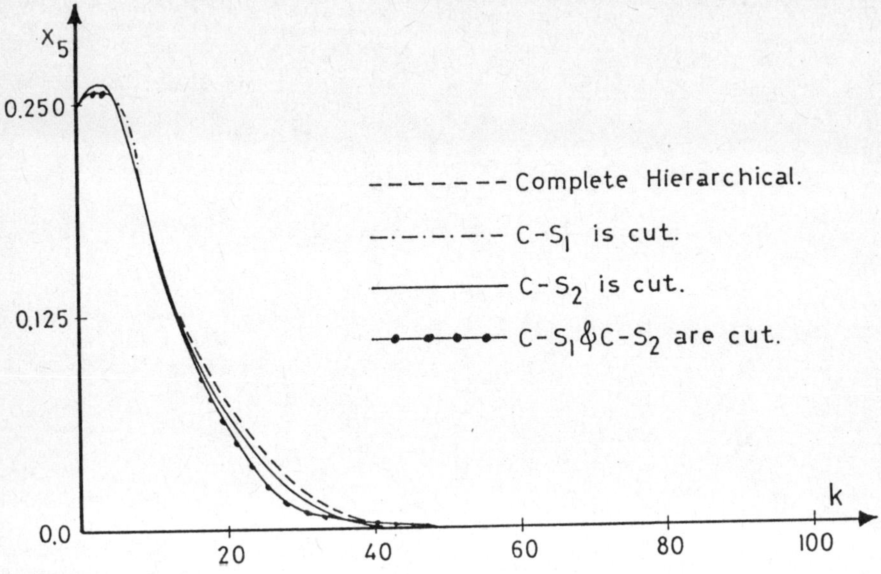

Figure 7.15

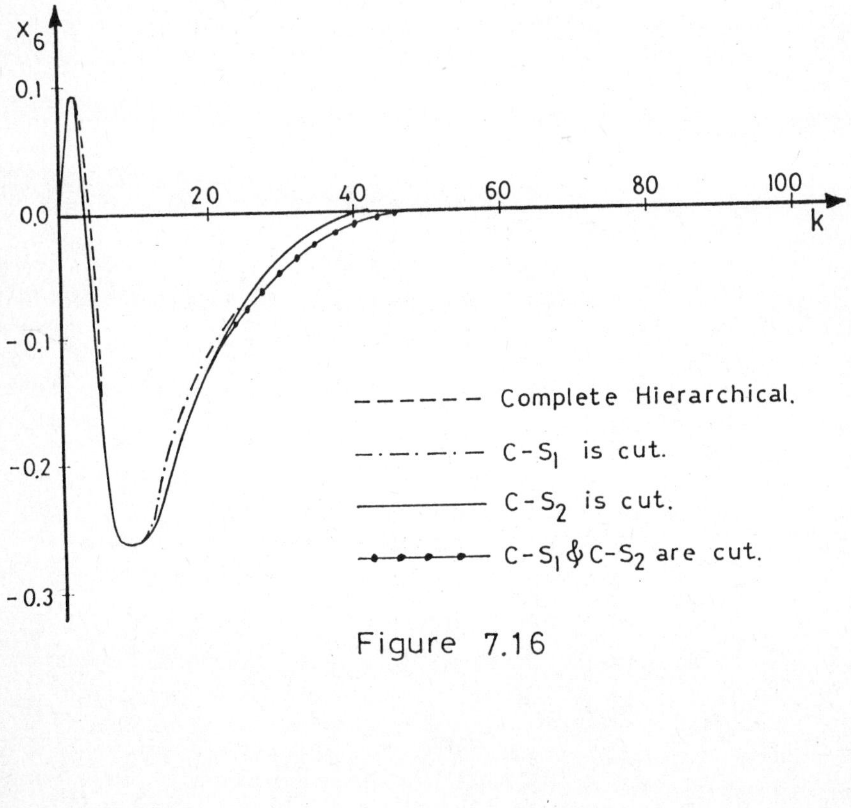

Figure 7.16

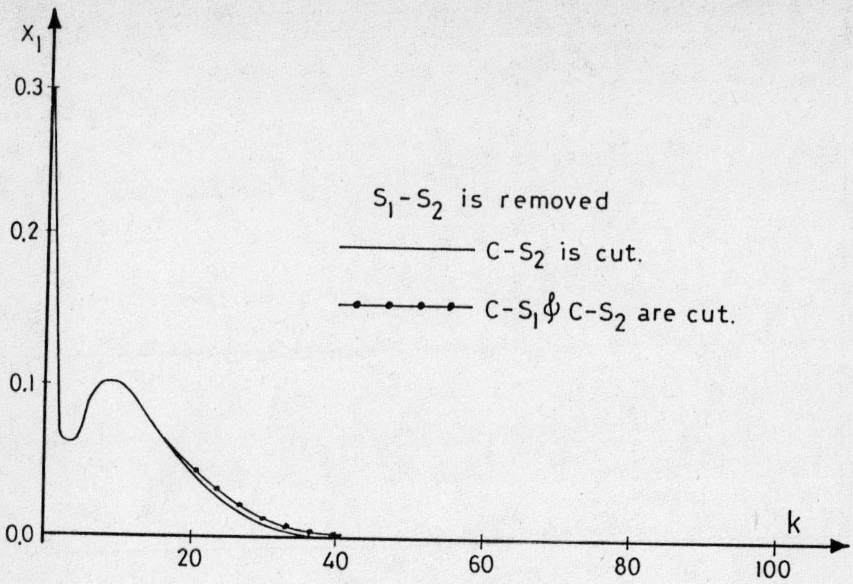

Figure 7.17

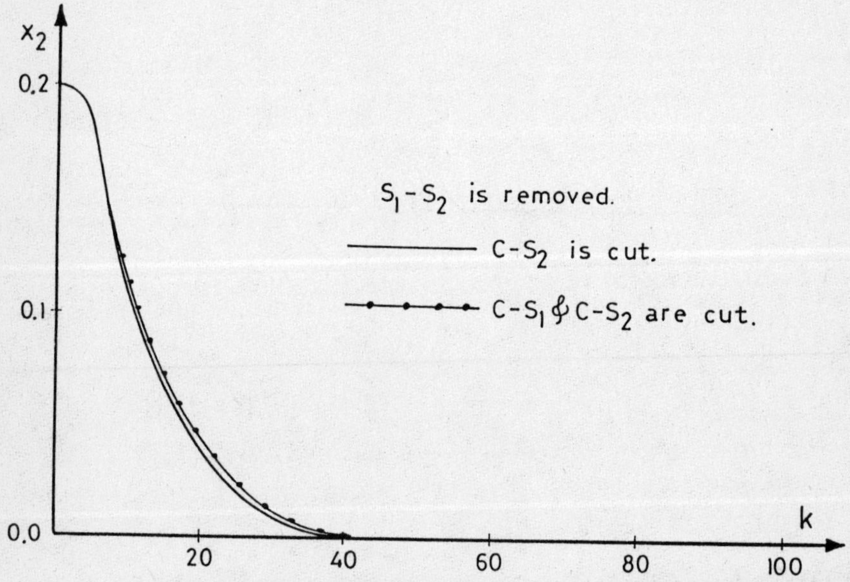

Figure 7.18

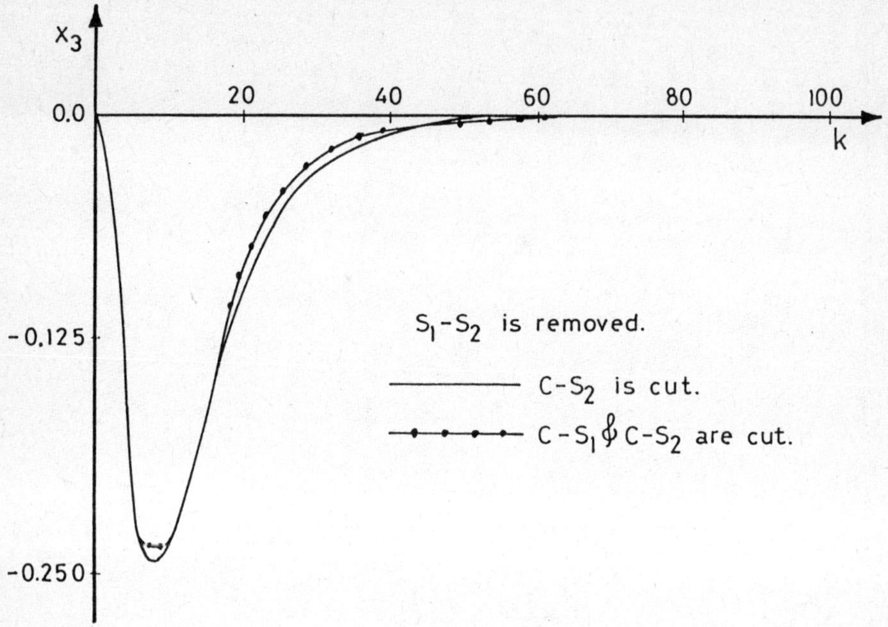

Figure 7.19

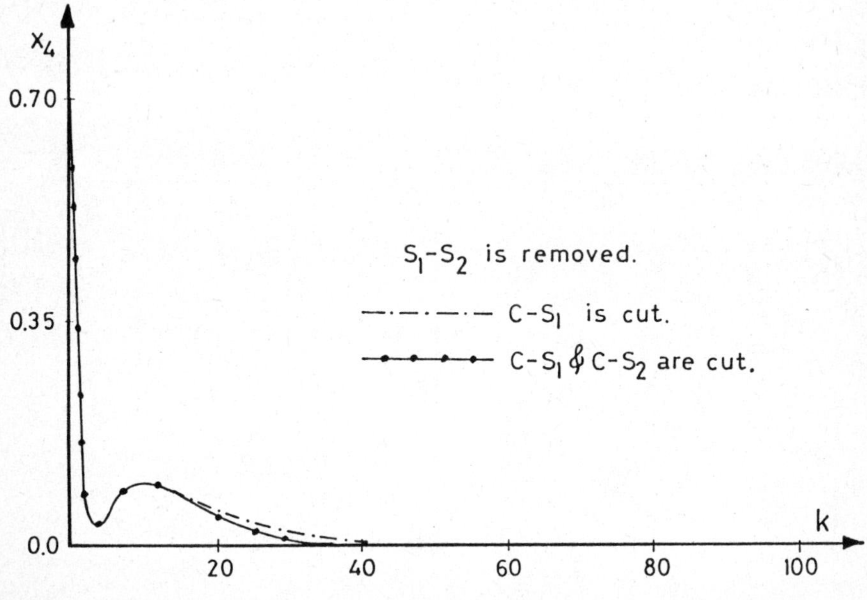

Figure 7.20

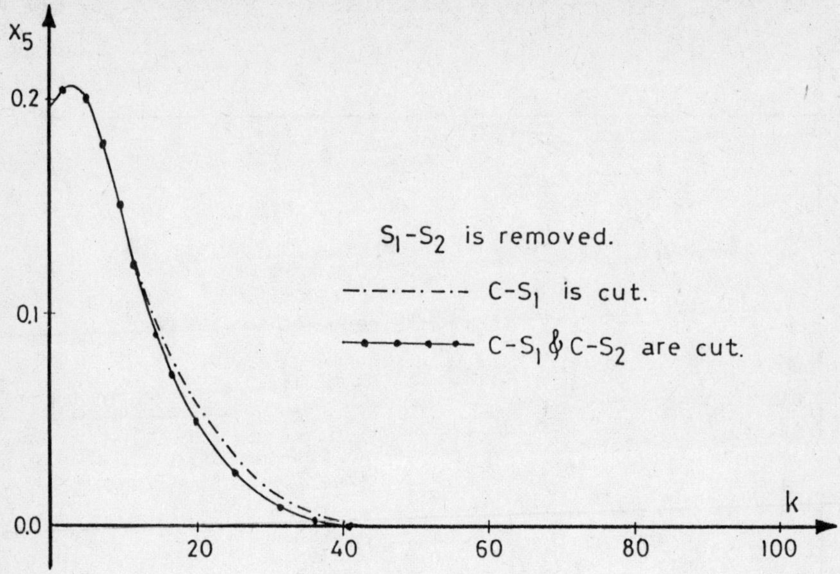

Figure 7.21

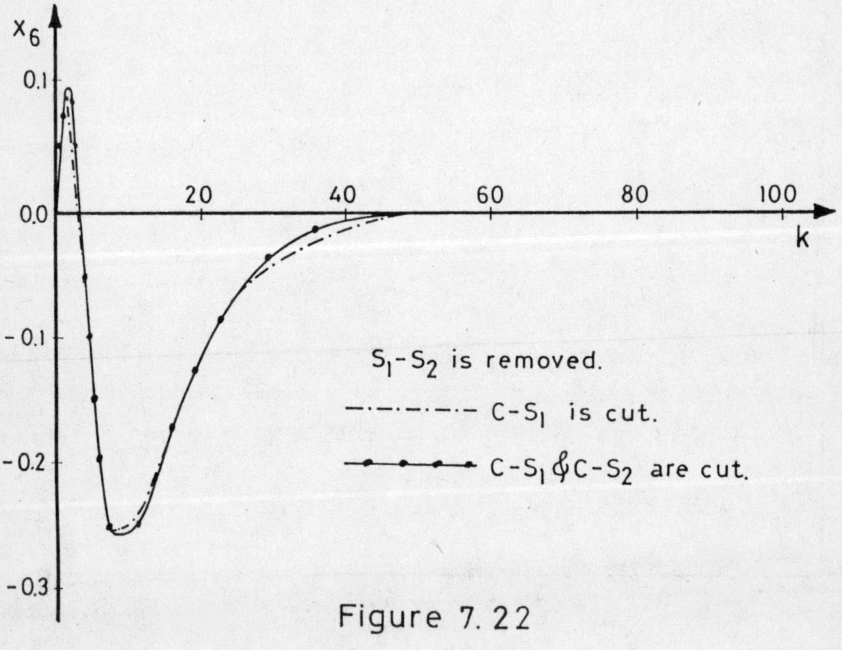

Figure 7.22

The resulting closed-loop state trajectories are shown in Figures (7.11-7.16). We performed the following structural perturbations between the coordinator and subsystems :

(a) link (C - S_1) is cut , (b) link (C - S_2) is cut and

(c) the two links (C - S_1) and (C - S_2) are removed.

For the purpose of comparison , the above three cases are plotted on the same graphs depicted in Figures (7.11-7.16).

Further simulation study is performed when the subsystems are decoupled, i.e. S_1-S_2 in Fig. 7.10 is removed. Moreover, the three-cases a,b,c are taken into consideration and the closed-loop state trajectories are plotted in Figs (7.17-7.22).

From these simulations , one can conclude that the resulted closed-loop uncertain system in each case is stable.

7.8 Conclusions

The purpose of this chapter has been to develop control schemes for large-scale interconnected, uncertain dynamical systems. It has been shown, from the theoretical and practical viewpoints, that the developed controllers are efficient in regulating the system operation, towards steady state, and adapting its performance against the bounded uncertainties. Moreover, the boundedness of the global system under various structural perturbations either in the interconnections between the subsystems and/or in the information network, is studied. From the simulation of the cited example, with all possible perturbations, it is clear that the system is always uniformly bounded.

CHAPTER 8

SUMMARY AND CONCLUSIONS

The main objective of the work conducted in this book is to develop efficient control techniques in order to stabilize different classes of partially-known dynamical systems. Particularly, focus is given to the important problems of achieving boundedness behaviour of the system at hand against internal bounded uncertainties and/or external bounded disturbances.

The main contributions of this book can be summarized as follows :

(1) The effect of unmodeled high frequency parasitics and/or external bounded disturbances on the stability and performance of reduced-order adaptive control schemes are analyzed in Chapters 2 and 3 for continuous and discrete-time systems respectively. Many adaptation mechanisms are developed in this respect. It has been shown that for uniformly bounded input, reduced-order controllers yield stable performance but within a prescribed region of attraction. This region contains all signals which will converge to target sets around the equilibrium for perfect adaptive regulation and tracking. The size of the target sets depend on the adaptation gain, the mode-separation ratio, the adaptation mechanism, the external bounded disturbances and the magnitude and priodicity of the reference input.

(2) In Chapter 4, we considered nominally linear time-invariant systems which contain uncertain elements and are subject to uncertain inputs, and for which either uncertain state or output is available. We constructed feedback controls (linear and/or nonlinear) utilizing measured states,

if available, or measured output. It has been shown that the developed controllers, under the validity of the matching conditions, either asymptotic stability or ultimate boundedness within a certain neighborhood of the zero state, no matter what the actual evolution of the uncertain quantities is. It should emphasized that if state error is sufficiently large, the nonlinear control may be "worse" than linear control, in that the magnitude of the former may exceed that of the latter for the same guaranteed region of ultimate boundedness. However, as state error decreases, nonlinear control becomes more and more efficacious vis-a-vis linear control. Eventually, for zero state error, nonlinear control guarantees asymptotic stability whereas linear control of any bounded magnitude does not.

(3) In Chapter 5, the problem of stabilizing a class of uncertain linear systems, in the absence of matching conditions is addressed. A two-level control scheme is developed which can maintain the optimal behaviour of the system under normal operating conditions and at the same time guarantee both reliability and stability of the system for a wide class of model uncertainty.

(4) A general class of discrete-time systems having time varying "unknown but bounded" parameters with incomplete state measurements, have been presented in section 5.4. It has been shown that robust stabilization of the uncertain system via observers is possible provided that the uncertain parameters don't exceed a certain computable threshold.

(5) New decentralized and hierarchical control schemes are developed for interconnected uncertain, continuous and discrete-time systems in Chapters 6 and 7 respectively. These strategies are shown to be compatible and resulted in a stable overall system. Furthermore, The closed-loop system is proved to be "connectively bounded", i.e. the overall system

is bounded under some structural perturbations. In addition to this, it has been shown that the controllers'structures don't need transfer of information between the subsystems.

From the work undertaken and the experience gained, one can propose several open problems, in the area of uncertain dynamical systems, which can be investigated in the future.

(i) In the area of reduced-order adaptive control, one difficulty is that the analysis to date has only been applied to the simplest model reference adaptive control algorithm using a gradient parameter estimator. Inherent limitations are present. The model reference controller is fundamentally non-robust (even when the true parameters are known) since it incorporates feedforward terms. Also, the gradient procedure is known to converge slowly. Thus, it would seem that more work is needed on the robustness question from two points of view. Firstly, attention should be given to adaptive (versions of) robust control for the algorithms. Then, these algorithms could be subjected to robustness along the singular perturbation analysis.

(ii) In the development of nonlinear control for a class of uncertain discrete systems described in Chapter 4, one may argue that approximating a relay action by a linear saturated control is not allowed in practice, since in the neighborhood of the switching surface, the action is completely unpredictable. The relaxation of this approximation is generally possible and promising for future research.

(iii) The extension of the developed two-level control structure in Chapter 5 to deal with more general class of uncertain systems in the absence of matching conditions, seems to be interesting. Moreover, the study of other stabilization schemes associated with these systems is a challenging target.

(iv) In Chapter 5, sufficient conditions were given for separate design of controller and observer for a general class of uncertain dynamical systems. In this area of future research, it would be of interest to focus attention on the optimal selection of the observer gain matrix " M ". It should be emphasized that we considered a one-step delay in the measurement pattern. Another area for future research would involve the situation where the outputs are recorded instantaneously.

(v) Another interesting problem is to extend the analytical framework (in Chapters 6 and 7), based on what is called "overlapping decomposition idea", [89,99,100] in order to obtain near optimal decentralized feedback control strategies for interconnected uncertain dynamical systems.

(vi) Decentralized stabilization of large-scale uncertain dynamical systems is an exciting open problem for future research. In this area a number of promising results have been obtained in [95-97,141-144].

REFERENCES

[1] Astrom, K. and B. Wittenmark ,
"Computer Controlled Systems : Theory and Design" ,
Prentice - Hall , Inc. , Englewood Cliffs , N.J. , 1984.

[2] Mahmoud , M. S. and M. G. Singh ,
"Discrete Systems: Analysis , Optimization and Control",
Springer - Verlag , Berlin , 1984.

[3] Aoki , M. ,
"Optimization of Stochastic Systems",
New York : Academic , 1967.

[4] Landau , I. D.,
"Adaptive Control : The Model Reference Approach",
New York , Marcel Dekker , 1979.

[5] Mahmoud , M. S.,
"Computational Experience With Adaptive Model - Reference Identification Schemes" , Int. J. Systems Science , Vol. 18 , No. 3 , PP. 527-542 , 1987.

[6] M. Sobral , Jr. ,
"Senesitivity in Optimal Control Systems" ,Proc. IEEE ,
Vol. 56 , pp. 1644-1652 , Oct. 1968.

[7] Werner, R. A. and J. B. Cruz, Jr.,
"Feedback Control Which Preserves Optimality for Systems With Unknown Parameters",IEEE Trans. Automatic Control,
Vol. AC-13, pp. 621-629, Dec. 1968.

[8] Kokotovic,P.V., J.B. Cruz,Jr., J.E.Heller and P.Sannuti,
"Synthesis of Optimally Sensitive Systems", Proc. IEE,
Vol. 56, pp. 1318-1324, Aug. 1968.

[9] Chang , S, S. L. and T. K. C. Peng ,
"Adaptive Guaranteed Cost Control of Systems With Uncertain Parameters" , IEEE Trans. on Automatic Control , Vol. AC - 17 , No. 4 , pp. 474-483 , Aug. 1972

[10] Roxin , E. ,
"On Generalized Dynamical Systems Defined by a Contingent Equation" , Journal of Differential Equations , Vol. 1 , No. 2 , 1965

[11] Salmon , D. ,
"Minimax Controller Design" , IEEE Trans. on Automatic Control , Vol. AC - 13 , pp. 369-376 , Aug. 1968.

[12] Gutman , S.
"Uncertain Dynamical Systems - A Lyapunov Min-Max Approach" ,IEEE Trans. on Automatic Control , Vol. 24 , No. 3 , PP 437-443 , June 1979.

[13] Narendra, K. S. and P. Kudva,
" Stable Adaptive Schemes For System Identification and Control-Parts I&II",IEEE Trans. System, Man and Cybernitics, Vol. SMC-4, pp. 541-460, Nov. 1974.

[14] Narendra, K. S. and L. S. Valvani,
"Stable Adaptive Controller Design-Direct Control",IEEE Trans. Automatic Control, Vol. AC-23, pp. 570-583, 1978.

[15] Narendra, K. S., Y. H. Lin and L. S. Valvani,
" Stable Adaptive Controller Design, Part II: Proof of Stability", Vol. AC-25, pp. 440-448, 1980.

[16] Narendra, K. S. and Y. H. Lin
"Stable Discrete Adaptive Control",IEEE Trans. Automatic Control, Vol. AC-25, pp. 456-461, 1980.

[17] Goodwin, G. C., P. J. Ramadge and P. E. Chains, " Discrete-Time Multivariable Adaptive Control", IEEE Trans. Automatic Control, Vol. AC-25,, pp.449-456, 1980.

[18] Goodwin, G. C. and K. S. Sin, "Adaptive Control of Nonminimum Phase Systems", IEEE Trans. Automatic Control, Vol. AC-26, pp. 478-483,1981.

[19] Goodwin, G. C., D. H. Hill and M. Palaniswami, " A Perspective on Convergence of Adaptive Control Algorithms",Automatica, Vol.20, pp.519-531, 1984.

[20] Goodwin, G. C. and K. S. Sin, "Adaptive Filtering, Prediction and Control", Englewood Cliffs, New Jersy, Prentice-Hall, 1984.

[21] Kreisselmeier, G., "On Adaptive State Regulation", IEEE Trans. Automatic Control, Vol. AC-27, pp. 3-17, 1982.

[22] Elliot, H., R. Cristi and M. Das, "Global Stability of Adaptive Pole Placement Algorithms" IEEE Trans. Automatic Control, Vol. AC-30, pp. 348-356, 1985

[23] Anderson, B.D.O. and R. M. Johnson, "Global Adaptive Pole Positioning",IEEE Trans. Automatic Control, Vol. AC-30, No. 1, pp. 11-22, 1985.

[24] Kreisselmeier, G. and B.D.O. Anderson, "Robust Model Reference Adaptive Control",IEEE Trans. Automatic Control, Vol. AC-31,No. 11, pp. 127-133,1986.

[25] Rohrs, C., L. Valvani, M. Athans and G. Stain, " Analytical Verification of Undesirable Properties of Direct Model Reference Control Algorithms", 20 th IEEE

Conf. On Decision and Control, San Diego, CA, pp. 1272-1284, 1981.

[26] C. Rohrs, L. Valvani, M. Athans and G. Stain, "Robustness of Adaptive Control Algorithms in the Presence of Unmodeled Dynamics", 21 th IEEE Conf. On Decision and Control, Orlando, FL, Dec. 1982.

[27] Egardt, B., "Stability of Adaptive Controllers", Lectures Notes in Control and Information Sciences, Springer-Verlag, Berlin , 1979.

[28] Peterson, B. B. and K. S. Narendra, "Bounded Error Adaptive Control", IEEE Trans. Automatic Control, Vol. AC-27, No.6, pp. 1161-1168, 1982.

[29] Kreisselmeir, G. and K. S. Narendra, "Stable Model Reference Adaptive Control in the Presence of Bounded Disturbances", IEEE Trans. Automatic Control, Vol. AC-27, No. 6, pp.1169-1175, 1982.

[30] Samson, C., "Stability Analysis of Adaptively Controlled System Subject to Bounded Disturbances", Automatica, Vol. 19, pp.81-86, 1983.

[31] Narendra, K. S. and A. M. Annaswamy, " Robust Adaptive Control in the Presence of Bounded Disturbance", IEEE Trans. Automatic Control, Vol. AC-31 , No. 4, pp. 306-315, April 1986.

[32] Ioannou, P. A. and P. V. Kokotovic, "Adaptive Systems With Reduced Models", Lectures Notes in Control and Information Sciences, Springer-Verlag, Berlin, 1983.

[33] Ioannou, P. A. and K. S. Tsakalis, " A Robust Direct Adaptive Controller", IEEE Trans. Automatic Control, Vol. AC-31, No.11,pp.1033-1043, 1986.

[34] Ioannou, P. A. and P. V. Kokotovic, "Robust Redesign of Adaptive Control", IEEE Trans. Automatic Control, Vol. AC-29, No. 3, pp.202-212, 1984.

[35] Ioannou, P. A. and P. V. Kokotovic, "Instability Analysis and Improvement of Robustness of Adaptive Control",Automatica, Vol. 20, No. 5, pp. 583-594, 1984.

[36] Ioannou, P. A., "Robust Adaptive Controller With Zero Tracking Errors", IEEE Trans. Automatic Control, Vol. AC-31, No. 8, pp. 773-776, 1986.

[37] Vinkler, A. and I. J. Wood, " Multistep Guaranteed Cost Control of Linear Systems With Uncertain Parameters", J. Guidance and Control, Vol. 2, No. 6, pp. 449-456, 1980.

[38] Leitmann , G. " Guaranteed Ultimate Boundedness For A Class of Uncertain Linear Dynamical Systems " , IEEE Trans. Automatic Control, Vol. AC-23, No. 6,pp.1109-1110, 1978.

[39] Leitmann , G. " Guaranteed Asymptotic Stability For Some Linear Systems With Bounded Uncertainties",ASME Journal of Dynamic Systems, Measurement and Control, Vol. 101, No. 3, pp. 212-219 , Sept. 1979.

[40] Leitmann , G. "On the Efficacy of Nonlinear Control in Uncertain

Linear Systems " , ASME Journal of Dynamic Systems , Measurement and Control, Vol. 102 , No.2 , pp. 95-102 , June 1981.

[41] Corless , M. J. and G. Leitmann ,
" Continuous State Feedback Guaranteeing Uniform Ultimate Boundedness For Uncertain Dynamic Systems " , IEEE Trans. Automatic Control, Vol. AC-26, No. 5,pp. 1139-1144,1981.

[42] Barmish , B. R. and G. Leitmann ,
" On Ultimate Boundedness Control of Uncertain Systems in the Absence of Matching Conditions " , IEEE Trans. Automatic Control, Vol. AC-27, No. 1, pp. 153-158, 1982.

[43] Aizerman , M. A. and Ye. S. Piatnitskiy ,
" Theory of Dynamic Systems Which Incorporate Elements With Incomplete Information and its Relation to the Theory of Discontinuous Systems " , Journal of the Franklin Institute of Technology , Vol. 306 , No. 6 , pp. 379-408, 1978.

[44] Petersen , I. R. ,
" Quadratic Stabilizability of Uncertain Linear Systems: Existence of A Nonlinear Stabilizing Control Does not Imply Existence of Linear Stabilizing Control " , IEEE Trans. Automatic Control , Vol. AC-30 , No. 3 , pp. 291-293, March 1985.

[45] Barmish , B. R. , M. Corless and G. Leitmann ,
" A New Class of Stabilizing Controllers For Uncertain Dynamical Systems " , SIAM J. Control and Optimaization , Vol. 21 , No.2 , pp. 246-255 , March 1983.

[46] Barmish , B. R. ,
"Stabilization of Uncertain Systems via Linear Control", IEEE Trans Automatic Control , Vol. AC-28 , No. 8 , pp. 848-850 , Aug. 1983.

[47] Barmish, B. R. ,I. R. Petersen and A. Feuer ,
" Linear Ultimate Boundedness Control of Uncertain Dynamical Systems " , Automatica , Vol. 19 , No. 5 , pp. 523-532 , 1983.

[48] Petersen , I. R. ,
" A Riccati Equation Approach to the Design of Stabilizing Controllers and Observers for a Class of Uncertain Linear Systems " , IEEE Trans. Automatic Control , Vol. AC-30 , No. 9 , pp. 904-907 , Sept. 1985.

[49] Barmish , B. R. ,
" Necessary and Sufficient Conditions for Quadratic Stabilizability on an Uncertain Linear Systems", J. Opt. Theory and Applications, Vol. 46, pp. 399-408, 1985.

[50] Petersen , I. R. ,
" Structural Stabilization of Uncertain Systems : Necessity of the Matching Conditions " , SIAM J. Control, Vol. 23, pp. 286-296, 1985.

[51] Petersen , I. R. and C. V. Hollot ,
" A Riccati Equation Approach to the Stabilization of Uncertain Linear Systems " , Automatica , Vol. 22 , No. 4 , pp.397-411 , 1986.

[52] Astrom, K. J.,
"Adaptive Feedback Control", Proc. of the IEEE, Vol. 75, No. 2, pp. 185-215, Feb. 1987.

[53] Egardt, B.,
" Unification of Some Discrete-Time Adaptive Control Schemes", IEEE Trans. Automatic Control, Vol. AC-25, pp. 693-697, 1980.

[54] Anderson, B. D. O. and C. R. Johnson, Jr., "Exponential Convergence of Adaptive Identification and Control Algorithms", Automatica, Vol.18, pp.1-13, 1982.

[55] Mahmoud, M. S., H. A. Othman and N. M. Khraishi, "Reduced-Order Performance of Adaptive Control Systems", IEEE Trans. Automatic Control, Vol. AC-31, No. 11, pp. 1076-1079, 1986.

[56] Anderson, B. D. O. , "Adaptive Systems, Lack of Persistency of Excitation and Bursting Phenomena", Automatica, Vol. 21, No.3, 1985.

[57] Hsu, L. and R. R. Costa, "Bursting Phenomena in Continuous-time Adaptive Systems With a σ-modification", IEEE Trans. Automatic Control, Vol. AC-32, No. 1, pp. 84-86, Jan. 1987.

[58] Narendra, K. S. and A. M. Annaswamy, "A New Adaptive Law for Robust Adaptation Without Persistent Excitation ",IEEE Trans. Automatic Control, Vol. AC-32, No.2, pp.134-145, Feb. 1987.

[59] Mahmoud, M. S. and A. A. Bahnasawi, "Stabilization of Discrete Adaptive Systems With Bounded Disturbances",Int. J. System Sciences,Vol. 19,Dec. 1988.

[60] Phillips, R. G., "Reduced-Order Modeling and Control of Two-Time-Scale Discrete Systems", Int. J. Contr., Vol. 31, pp. 765-780, 1980.

[61] Kokotovic,P.V., J.J.Allemong, J.R.Winkelman and J.H.Chow "Singular Perturbation and Iterative Separation of Time Scales", Automatica, Vol. 16, pp. 23-31, 1980.

[62] Mahmoud, M. S.,
"Structural Properties of Discrete Systems with Slow and Fast Modes", Large Scale Systems, Vol. 3, pp. 227-236, 1982.

[63] Mahmoud, M. S.,
"Order Reduction and Control of Discrete Systems", Proc. IEE, Vol. 129, pp. 129-135, 1982.

[64] Saksena, V. R., J. O'Reilly and P. V. Kokotovic,
"Singular Perturbations and Time-Scale Methods in Control Theory: Survey 1976-1983", Automatica, Vol. 20, No. 3, pp. 273-293, 1984.

[65] Mahmoud, M. S. and A. A. Bahnasawi,
" Some Properties of Absolutely-Stable Discrete-Time Systems ", J. Control Theory and Advanced Technology, Vol. 4, No. 2, pp. 243-250, June 1988.

[66] Bahnasawi, A. A. and M. S. Mahmoud,
"Reduced-Order Adaptive Control Scheme", (Submitted for publication).

[67] Bahnasawi, A. A. and M. S. Mahmoud,
" Robust Control of Reduced-Order Adaptive-Discrete Systems",Control Theory and Advanced Technology,Submitted

[68] Mahmoud, M. S. and A. A. Bahnasawi,
" Asymptotic Stability for a Class of Linear Discrete Systems With Bounded Uncertainty", IEEE Trans. Automatic Control, Vol. AC-33, No. 6, pp.572-575, June 1988.

[69] Bahnasawi, A. A. and M. S. Mahmoud,
"Uncertain Discrete Systems : Uniform Ultimate Bounded Stabilization", Int. J. of Systems Sciences, Submitted.

[70] Bahnasawi, A. A. , Al-Fuhaid, A. S. and M. S. Mahmoud , "Linear Feedback Approach to the Stabilization of Uncertain Discrete Systems", Proc. IEE Part D, to appear.

[71] Berge, C., " Topological Spaces", London : Oliver and Boyd, 1963.

[72] Kalman , R. E. and J. E. Bertram , " Control System Analysis and Design via the 'Second Method of Lyapunov', I. Continuous - Time Systems , II. Discrete - Time Systems " , Trans. ASME J. Basic Engineering , Vol. 82 D, pp. 371-400 , June 1960.

[73] Vidyasagar , M. " Nonlinear Systems Analysis " , Prentice-Hall , Inc. , Englewood Cliffs , N. J. , 1978.

[74] Vidyasagar, M. , "New Direction of Research in Nonlinear System Theory" , Proc. IEEE , Vol. 74, No. 8, pp. 1060-1091, Aug. 1986.

[75] Bahnasawi, A. A. and M. S. Mahmoud, "Stabilization of Uncertain Discrete Systems via Two-level Controller Structures", 12th IMACS World Congress on Scientific Computation, Paris, FRANCE, pp. 246 - 248, July 18-22, 1988.

[76] Kwakernaak, H. and R. Sivan, " Linear Optimal Control", Wiley, New York, 1972.

[77] Luenberger, D. G., " Observers For Multivariable Systems", IEEE Trans. Automatic Control, Vol. AC-11, pp. 190-197, 1966.

[78] O'Reilly, J., "Observers For Linear Systems", Academic Press, Inc., London, 1983.

[79] Mahmoud, M. S.,
" Design of Observer-Based Controller For A Class of Discrete Systems",Automatica, Vol.18, pp. 323-328, 1982.

[80] Willems, J. L.,
"Design of State Observers For Linear Discrete Systems", Int. J. Systems Science, Vol. 11, pp. 139-147, 1980.

[81] Thau, F. E. and A. Kestenbaum,
" The Effect of Modeling Errors of Linear State Reconstructors and Regulators", ASME J. of Dynamic System, Measurement and Control, Vol. 96, pp. 454-459, 1974.

[82] Barmish , B. R. and A. R. Galmidi ,
" Robustness of Luenberger Observers : Linear Systems Stabilized via Nonlinear Control " , Automatica , Vol. 22 , No. 4 , pp. 413-423 , 1986.

[83] Bahnasawi, A. A. and M. S. Mahmoud,
" Robust Stabilizing Observer-Based Controller For Unknown Discrete Systems", 4th IFAC Symposium on Computer Aided Design in Control Systems (CADCS'88), Beijing, P.R. CHINA, Aug. 23-25, 1988.

[84] Albert, A. A.,
" Regression and The Moore-Penrose Pseudo Inverse", Academic Press, New York, 1972.

[85] Hassan , M. F. , A. Titli , M. G. Singh and R. Hurteau ,
" Stability , Stabilization and Performance of Multilevel Controllers Under Structural Perturbations , I. Cutting the Links Between Co-ordinator and Subsystems " , IEE Proc., Vol. 127, Pt. D, No. 5, pp. 207-213, Sept. 1980.

[86] Hassan , M. F. and M. G. Singh ,
" Stability , Stabilization and Performance of Multilevel Controllers Under Structural Perturbations , II. Stabilization Under any Structural Perturbations " , IEE Proc. , Vol. 127 , Pt. D , No. 5 , pp. 214-219 , Sept. 1980.

[87] Hassan , M. F. , A. A. Bahnasawi and S. Z. Eid ,
" A Hierarchical Approach For the Megawatt - Frequency Control of Multi - Area Power Systems " , 11 th IMACS World Congress , Oslo , Norway , Aug. 5-9 , 1985.

[88] Singh , M. g. and A. Titli ,
" Systems : Decomposition , Optimization and Control " , Pergamon Press , Oxford , 1978.

[89] Siljak , D. D.
" Large-Scale Dynamic Systems : Stability and Structure " , North - Holland , Inc. , New York , 1978.

[90] Mahmoud , M. S. and M. G. Singh ,
" Large - Scale Systems Modeling " , Pergamon Press , Oxford , 1981.

[91] Jamshidi , M.
" Large - Scale Systems : Modeling and Control " , North - Holand , Inc. , New York , 1983.

[92] Mahmoud , M. S. , M. F. Hassan and M. G. Darwish ,
" Large Scale Dynamic Systems : Theories and Techniques " , Marcel Dekker , Inc. , New York , 1985.

[93] Anderson , B. D. and J. B. Moore ,
" Linear Optimal Control " , Prentice - Hall , Inc. , Englewood Cliffs , N. J. , 1971.

[94] Lapidus , L. and N. R. Amundson ,
" Stagewise Absorbation and Extraction Equipment : Transient and Unsteady State Operation " , Ind. Eng. Chem. , Vol. 42 , pp. 1071-1078 , 1950.

[95] Bahnasawi, A. A., Al-Fuhaid, A. S. and M. S. Mahmoud,
" Decentralized and Hierarchical Control of Interconnected Uncertain Systems",Proc. IEE Part D, Submitted.

[96] Bahnasawi, A. A., Al-Fuhaid, A. S. and M. S. Mahmoud,
" Stabilization of Interconnected Uncertain Discrete Systems via Decentralized Control", Information and Decision Technologies, to appear.

[97] Bahnasawi, A. A., Al-Fuhaid, A. S. and M. S. Mahmoud,
" A New Hierarchical Control Structure For A Class of Uncertain Discrete Systems", Control Theory and Advanced Technology, Submitted.

[98] Zohdy, M. A., N. K. Loh and A. Abdul-Wahab,
" A Robust Optimal Model Matching Control ", IEEE Trans. Automatic Control, Vol. AC-32, No. 5, pp. 410-414, 1987.

[99] Bahnasawi, A. A., M. F. Hassan and S. Z. Eid,
"A New Decentralized Controller For The Interconnected Egyptian Power Network", Large Scale Systems Theory and Application, Vol. 11, pp. 217-232, 1986.

[100] Hassan, M. F., A. A. Bahnasawi and R. H. Ragab,
"A New Overlapped Decentralized Controller For Large Scale Systems",Int. J. Systems Science, Vol. 18, No. 10 pp. 1963-1977, 1987.

[101] Chen, Y. H. and G. Leitmann,
"Robustness of Uncertain Systems in the Absence of Matching Assumptions", Int. J. Control,45, 1527, 1987.

[102] Thorp, J. S. and B. R. Barmish,
"On Guaranteed Stability of Uncertain Systems via Linear Control", J. Optimiz. Theory Appl., 35, 559, 1981.

[103] Stalford, H. L.,
"Robust Control of Uncertain Systems in the Absence of Matching Conditions : Scalar Input", Proc. Conf. Decision and Control, 1987.

[104] Corless, M. and G. Leitmann,
"Adaptive Control of Systems Containing Uncertain Functions and Unknown Functions With Uncertain Bounds", J. Optimiz. Theory Appl., 41, 155, 1983.

[105] Corless, M. and G. Leitmann,
" Adaptive Control of Uncertain Dynamical Systems", in Dynamical Systems and Microphics : Control Theory and Mechanics, (A. Blaquiere and G. Leitmann, eds.), Academic Press, New York, 1984.

[106] Corless, M.,
" Controlling Uncertain Systems Within a Subset of the State Space", Proc. American Control Conf., Boston, 1985.

[107] Corless, M., G. Leitmann and J. M. Skowronski,
"Adaptive Control for Avoidance or Evasion in an Uncertain Environment", Computers and Mathematics With Applications, 13, 1, 1987.

[108] Leitmann, G.,
"Guaranteed Avoidance Strategies", J. Optimiz. Theory Appl., 32, 569, 1980.

[109] Leitmann, G. and J. Skowronski,
"Avoidance Control", J. Optimiz. Theory Appl., 29, 581, 1977.

[110] Leitmann, G. and J. Skowronski,
"A Note on Avoidance Control", Optimal Control Appl. Methods, 4, 335, 1983.

[111] Thowsen, A.,
"Uniform Ultimate Boundedness of the Solutions of the Uncertain Dynamic Delay Systems with State-Dependent and Memoryless Feedback Control", Int J. Control, 37, 1135, 1983.

[112] Yu, Y.,
" On Stabilizing Uncertain Linear Delay Systems", J. Optimiz. Theory Appl., 3, 503, 1983.

[113] Corless, M.
"Stabilization of Uncertain Discrete-Time Systems",Proc. IFAC Workshop on Model Error Concepts and Compansation, Boston, 1985.

[114 Corless, M. and J. Manela,
"Control of Uncertain Discrete-Time Systems", Proc. American Conf., Seattle, Washington, June 18-20, 1986, pp. 515-520.

[115] Magana, M. E. and S. H. Zak,
"Robust State Feedback Stabilization of Discrete-Time Uncertain Dynamical Systems", IEEE Trans. Automatic Control, Vol. AC-33, No. 9, pp. 887-891, 1988.

[116] Corless, M.,
"Robustness of a Class of Feedback-Controlled Uncertain Nonlinear Systems in the Presence of Singular Perturbations", Proc. American Control Conf., pp. 1584-1589, Minneapolis, 1987.

[117] Leitmann, G. and E. P. Ryan,
"Output Feedback Control of a Class of Singularly Perturbed Uncertain Dynamical Systems", Proc. American Control Conf., pp. 1590-1594, Minneapolis, 1987.

[118] Leitmann, G., E. P. Ryan and A. Steinberg,
"Feedback Control of Uncertain Systems : Robustness with Respect to Neglected Actuator and Sensor Dynamics", Int. J. Control, 43, 1243, 1986.

[119] Garofalo, F. and G. Leitmann,
"Nonlinear Composite Control of a Nominally Linear Singularly Perturbed Uncertain System", Proc. 12th IMACS World Congress, Paris, FRANCE, 1988.

[120] Chen, Y. H.,
"Robust Output Feedback Controller : Direct Design", Int. J. Control, 46, 1083, 1987.

[121] Chen, Y. H.,
"Robust Output Feedback Controller : Indirect Design", Int. J. Control, 46, 1093, 1987.

[122] Galimidi, A. R. and B. R. Barmish,
"The Constrained Lyapunov Problem and its Application to Robust Output Feedback Stabilization" IEEE Trans. Automatic Control, Vol. AC-31, pp. 410-419, 1986.

[123] Steinberg, A. and M. Corless,
" Output Feedback Stabilization of Uncertain Dynamical Systems", IEEE Trans. Automatic Control, Vol. AC-30, pp. 1025-1027, 1985.

[124] Zeheb, E.,
"A Sufficient Condition for Output Feedback Stabilization of Uncertain Systems", IEEE Trans. Automatic Control Vol. AC-31, pp. 1055-1057, 1986.

[125] Walcott, B. L. and S. H. Zak,
"State Observation of Nonlinear Uncertain Dynamical Systems", IEEE Trans. Automatic Control, Vol. AC-32, pp. 166 - 170, 1987.

[126] Corless, M. and G. Leitmann,
"Controller Design For Uncertain Systems via Lyapunov Functions", Proc. of the 1988 American Control Conf., Atlanta, Georgia, 1988.

[127] Bremer, H. and A. Truckenbrodt,
"Robust Control for Industrial Robots", Proc. of RoManSy'84, Theory and Practice of Robots and Manipulators, MIT Press, 1985.

[128] Horowitz, R., H. I. Stephens, and G. Leitmann,
"Experimental Verification of a Deterministic Controller for a D. C. Motor with Uncertain Dynamics", Proc. American Control Conf., Minneapolis, 1987.

[129] Ryan, E. P., G. Leitmann and M. Corless,
"Practical Stabilizability of Uncertain Dynamical Systems, Application to Robotic Tracking", J. Optimiz. Theory Applic., 47, 235, 1985.

[130] Shoureshi, R., M. J. Corless and M. D. Roesler,
"Control of Industrial Manipulators with Bounded Uncertainties", J. Dynam. Syst. Meas. Contr., 109, 53, 1987.

[131] Gutman, S. and G. Leitmann,
"Stabilizing Feedback Control for Dynamical Systems with Bounded Uncertainty", Proc. IEEE Conf. Decision Control, 1976.

[132] Kelly, J., G. Leitmann and A. Soldatos, "Robust Control of Base-Isolated Structures Under Earthquake Excitation", J. Optimiz. Theory Appl., 52, 3, 1987.

[133] Singh, S. N., "Attitude Control of a Three Rotor Gyrostat in the Presence of Uncertainty", J. Astronautical Sciences, 35, 1987.

[134] Singh, S. N., "Nonlinear Adaptive Attitude Control of Spacecraft", IEEE Trans. Aerospace and Electronic Systems, Vol. AES-23, 371, 1987.

[135] Leitmann, G. and H. Y. Wan, Jr., "A Stabilization Policy for an Economy with Some Unknown Characteristics", J. Franklin Institute, 306, 23, 1978.

[136] Leitmann, G. and H. Y. Wan, Jr., "Macro-Economic Stabilization Policy for an Uncertain Dynamic Economy" in New Trends in Dynamic System Theory and Economics, Academic Press, New York, 1979.

[137] Leitmann, G. and H. Y. Wan, Jr., "Performance Improvement of Uncertain Macroeconomic Systems" in Dynamic Optimization and Mathematical Economics, (P.-T. Liu, ed), Plenum Press, New York, 1979.

[138] Chen, Y. H. and C. S. Lee, "On the Control of an Uncertain Water Quality System", Optimal Control Appl. Methods, 8, 279, 1987.

[139] Lee, C. S. and G. Leitmann, "Determinstic Control of an Uncertain Single Reach River A Discrete Model", Proc. of SICE'87, Hiroshima, Japan, 1987.

[140] Lee, C. S. and G. Leitmann,
"Uncertain Dynamical Systems : An Application to River Pollution Control", Proc. Modelling and Management of Resources Under Uncertainty, Honolulu, 1985, Lecture Notes in Biomathematics, 72, 167, Springer-Verlag, 1987.

[141] Leitmann, G., C. S. Lee and Y. H. Chen,
"Decentralized Control for a Large Scale Uncertain River System", Proc. IFAC Workshop on Modelling, Decisions and Games for Social Phenomena, Beijing, China, 1986.

[142] Leitmann, G., C. S. Lee and Y. H. Chen,
"Decentralized Control for an Uncertain Multi-Reach River System", Pro. Conf. on Optimal Control and Variational Calculus, Oberwolfach, 1986, West Germany, Lecture Notes in Control and Information Sciences, 95, Springer-Verlag, 1987.

[143] Chen, Y. H.,
" Deterministic Control of Large-Scale Uncertain Dynamical Systems", J. Franklin Institute, Vol. 323, No. 2, pp. 135-144, 1987.

[144] Chen, Y. H.,
"Decentralized Robust Control System Design for Large-Scale Uncertain Systems", Int. J. Control, Vol. 47, No. 5, pp. 1195-1205, 1988.

Lecture Notes in Control and Information Sciences

Edited by M. Thoma and A. Wyner

Vol. 81: Stochastic Optimization
Proceedings of the International Conference,
Kiew, 1984
Edited by I. Arkin, A. Shiraev, R. Wets
X, 754 pages, 1986.

Vol. 82: Analysis and Algorithms
of Optimization Problems
Edited by K. Malanowski, K. Mizukami
VIII, 240 pages, 1986.

Vol. 83: Analysis and Optimization
of Systems
Proceedings of the Seventh International
Conference of Analysis and Optimization
of Systems
Antiba, June 26-27, 1986
Edited by A. Bensoussan, J. L. Lions
XVI, 901 pages, 1986.

Vol. 84: System Modelling
and Optimization
Proceedings of the 12th IFIP Conference
Budapest, Hungary, September 2–6, 1985
Edited by A. Prékopa, J. Szelezsán, B. Strazicky
XII, 1046 pages, 1986.

Vol. 85: Stochastic Processes
in Underwater Acoustics
Edited by Charles R. Baker
V, 205 pages, 1986.

Vol. 86: Time Series and
Linear Systems
Edited by Sergio Bittanti
XVII, 243 pages, 1986.

Vol. 87: Recent Advances in
System Modelling and
Optimization
Proceedings of the IFIP-WG 7/1
Working Conference
Santiago, Chile, August 27-31, 1984
Edited by L. Contesse, R. Correa, A. Weintraub
IV, 199 pages, 1987.

Vol. 88: Bruce A. Francis
A Course in H_∞ Control Theory
XI, 156 pages, 1987.

Vol. 88: Bruce A. Francis
A Course in H_∞ Control Theory
X, 150 pages, 1987.
Corrected - 1st printing 1987

Vol. 89: G. K. H. Pang/A. G. J. McFarlane
An Expert System Approach to
Computer-Aided Design
of Multivariable Systems
XII, 223 pages, 1987.

Vol. 90: Singular Perturbations
and Asymptotic Analysis
in Control Systems
Edited by P. Kokotovic,
A. Bensoussan, G. Blankenship
VI, 419 pages, 1987.

Vol. 91 Stochastic Modelling
and Filtering
Proceedings of the IFIP-WG 7/1
Working Conference
Rome, Italy, Decembre 10-14, 1984
Edited by A. Germani
IV, 209 pages, 1987.

Vol. 92: L. T. Grujić, A. A. Martynyuk,
M. Ribbens-Pavella
Large-Scale Systems Stability Under
Structural and Singular Perturbations
XV, 366 pages, 1987.

Vol. 93: K. Malanowski
Stability of Solutions to Convex
Problems of Optimization
IX, 137 pages, 1987.

Vol. 94: H. Krishna
Computational Complexity
of Bilinear Forms
Algebraic Coding Theory and
Applications to Digital
Communication Systems
XVIII, 166 pages, 1987.

Vol. 95: Optimal Control
Proceedings of the Conference on
Optimal Control and Variational Calculus
Oberwolfach, West-Germany, June 15-21, 1986
Edited by R. Bulirsch, A. Miele, J. Stoer
and K. H. Well
XII, 321 pages, 1987.

Vol. 96: H. J. Engelbert/W. Schmidt
Stochastic Differential Systems
Proceedings of the IFIP-WG 7/1
Working Conference
Eisenach, GDR, April 6-13, 1986
XII, 381 pages, 1987.

Lecture Notes in Control and Information Sciences

Edited by M. Thoma and A. Wyner

Vol. 97: I. Lasiecka/R. Triggiani (Eds.)
Control Problems for Systems
Described by Partial Differential Equations
and Applications
Proceedings of the IFIP-WG 7.2
Working Conference
Gainesville, Florida, February 3-6, 1986
VIII, 400 pages, 1987.

Vol. 98: A. Aloneftis
Stochastic Adaptive Control
Results and Simulation
XII, 120 pages, 1987.

Vol. 99: S. P. Bhattacharyya
Robust Stabilization Against
Structured Perturbations
IX, 172 pages, 1987.

Vol. 100: J. P. Zolésio (Editor)
Boundary Control and Boundary Variations
Proceedings of the IFIP WG 7.2 Conference
Nice, France, June 10-13, 1987
IV, 398 pages, 1988.

Vol. 101: P. E. Crouch,
A. J. van der Schaft
Variational and Hamiltonian
Control Systems
IV, 121 pages, 1987.

Vol. 102: F. Kappel, K. Kunisch,
W. Schappacher (Eds.)
Distributed Parameter Systems
Proceedings of the 3rd International Conference
Vorau, Styria, July 6–12, 1986
VII, 343 pages, 1987.

Vol. 103: P. Varaiya, A. B. Kurzhanski (Eds.)
Discrete Event Systems:
Models and Applications
IIASA Conference
Sopron, Hungary, August 3-7, 1987
IX, 282 pages, 1988.

Vol. 104: J. S. Freudenberg/D. P. Looze
Frequency Domain Properties of Scalar
and Multivariable Feedback Systems
VIII, 281 pages, 1988.

Vol. 105: Ch. I. Byrnes/A. Kurzhanski (Eds.)
Modelling and Adaptive Control
Proceedings of the IIASA Conference
Sopron, Hungary, July 1986
V, 379 pages, 1988.

Vol. 106: R. R. Mohler (Editor)
Nonlinear Time Series and
Signal Processing
V, 143 pages. 1988.

Vol. 107: Y. T. Tsay, L.-S. Shieh, St. Barnett
Structural Analysis and Design
of Multivariable Systems
An Algebraic Approach
VIII, 208 pages, 1988.

Vol. 108: K. J. Reinschke
Multivariable Control
A Graph-theoretic Approach
274 pages, 1988.

Vol. 109: M. Vukobratović/R. Stojić
Modern Aircraft Flight Control
VI, 288 pages, 1988.

Vol. 110: In preparation

Vol. 111: A. Bensoussan, J. L. Lions (Eds.)
Analysis and Optimization
of Systems
XIV, 1175 pages, 1988.

Vol. 112: Vojislav Kecman
State-Space Models of Lumped
and Distributed Systems
IX, 280 pages, 1988

Vol. 113: M. Iri, K. Yajima (Eds.)
System Modelling and Optimization
Proceedings of the 13th IFIP Conference
Tokyo, Japan, Aug. 31 – Sept. 4, 1987
IX, 787 pages, 1988.

Vol. 114: A. Bermúdez (Editor)
Control of Partial Differential Equations
Proceedings of the IFIP WG 7.2
Working Conference
Santiago de Compostela, Spain, July 6–9, 1987
IX, 318 pages, 1989

Vol. 115: H.J. Zwart
Geometric Theory for Infinite
Dimensional Systems
VIII, 156 pages, 1989.

Vol. 116: M.D. Mesarovic, Y. Takahara
Abstract Systems Theory
VIII, 439 pages, 1989